THE BINARY UNIVERSE

Theoretical physics is driven by the ideas of the mathematical physicist and the sometimes, subjective interpretations of the mathematics of Relativity Theory and Quantum Mechanics. Yet today, we are still faced with many paradoxes and conundrums in our attempts to understand the laws of nature.

This visionary book stands back from the grindstone of mainstream physics and presents a down to earth, Engineer's view of Special and General Relativity, together with an insightful understanding of the physical nature of time that affects our view of the quantum world.

The author adheres strictly to causality throughout his deductive reasoning and, with just a little mathematics, arrives at an intuitive concept of the Grand Unification, Einstein's unfinished, unified field theory. The field provides a common source of energy for all processes (including gravitation), all matter from the beginning of the universe itself.

Ken Hughes demonstrates why we are living in a binary universe and his theory offers fascinating explanations for many present-day conundrums in physics.

Updated with even more explanations, examples and supporting research, this second edition of *The Binary Universe* revisits the conclusion of the first edition, with more detail and additional information. Contents include: Causality of gravitation, Wavelike time, Quantum time, Binary time, Entanglement and the quantum world. *The Binary Universe* opens up a radical new approach to Space and the universe that surrounds us.

Front Picture copyrights © rasica & © Sebastian Kaulitzki
Cover Design copyright © U P Publications 2018
Copyright Acknowledgements to the American National Aeronautics and Space Administration, NASA for photographic material used. Fonts: Caviar, Caviar Dreams © Lauren Thompson & Dustimo © 2002 Dustin Norlander

Copyright © 2014, 2018 Ken Hughes

Ken Hughes has asserted his moral rights

Published in Great Britain, 2018, by Hughes Publishing in co-operation with U P Publications, St George's Hse, George St, Huntingdon, Cambs, PE29 3GH. UK

A CIP Catalogue record of this book is available from the British Library

ISBN 978-0-9568002-4-4 Second Edition
(replaces ISBN 978-0-9568002-1-3)

2 7 0 8 6 4 3 5 9

Printed in England by The Lightning Source Group

www.uppbooks.com

2018

THE BINARY UNIVERSE

A Theory of Time

2nd Edition

Ken Hughes

ЧP

2018

"Time, among all concepts in the world of physics, puts up the greatest resistance to being dethroned from ideal continuum to the world of the discrete, of information, of bits....

Of all obstacles to a thoroughly penetrating account of existence, none looms up more dismayingly than 'time.'

Explain time?

Not without explaining existence.

Explain existence?

Not without explaining time.

To uncover the deep and hidden connection between time and existence... is a task for the future."

John Archibald Wheeler (1911 – 2008)

CONTENTS

ILLUSTRATIONS

Preface

"Today the network of relationships linking the human race to itself and to the rest of the biosphere is so complex that all aspects affect all others to an extraordinary degree. Someone should be studying the whole system, however crudely that has to be done, because no gluing together of partial studies of a complex nonlinear system can give a good idea of the behaviour of the whole."

Murray Gel Mann

This principle of how we should expand our knowledge of complex, non-linear systems clearly also applies to our attempts to understand the universe as a whole. Today, there are many ongoing and specialist avenues of investigation in mathematics and science with much effort targeted, simultaneously, in different independent directions. But, no one seems to have stepped back to take a view of the *whole system*, however crudely, to see how our current knowledge links together and describes the laws of nature in a fundamental way. This is what I have attempted to do, in this book.

In this, the second edition of "The Binary Universe", I have expanded some of the arguments and presented more supporting information. As a result, the theory is even more compelling and surely now, it demands attention from the leaders in mainstream physics. Thanks to the wealth of information made available since the last edition, I have updated some points and included some more speculative text, notably on the Big Bang, however, the basis of the theory remains unchanged. The underlying principles remain the same.

Despite having tested and otherwise challenged the theory against my expanding research since I first developed it, I have yet to find any fundamental flaws in the reasoning or anything in today's science that might contradict it. In fact, the more I research, the more the theory appears to be validated. It fits remarkably well with the latest research and I am now more convinced than ever that these ideas will form the basis for the science of the twenty-first century. There is no other view, which brings together so much of science at the logical, intuitive level than the Binary Universe Theory (B.U.T.), which also provides a common solution to many present-day conundrums.

As in the original edition, the first half of the book deals with the macro nature of time and shows how both Special and General Relativity can be interpreted purely as the effects of changing time rates. The notions of length contraction and space curvature can equally well be expressed as the effects of time rate change.

In the second half of the book, I deduce a radical and detailed definition for the nature of time so that, while Relativity Theory remains unaffected mathematically, a subtly different interpretation of its meaning emerges. Later, the book uses this new understanding to create and develop a new view of the world, a dynamic, wavelike, temporal understanding of our universe.

Mainstream scientists are currently stuck, pondering dimensions, shapes, and all things geometric. They are up a blind alley from which there is no escape unless they open their minds to the real, physical phenomenon of time and finally put their abstract geometry in its proper place. Those unwilling will have to be dragged, "kicking and screaming", into this new, temporal age of physics.

It is often said that the first step in scientific progress is to make a guess. Albert Einstein's initial guess for General Relativity for example was that space itself could actually curve and so he set about describing this mathematically. One should of course, make the best possible guess which conforms to known science and which seems the most likely to be true. It is of little value to follow a hunch or an idea that has no basis in accepted science. There would be no way of knowing if we were heading in the right direction and this would inhibit our attempts to prove the idea experimentally or mathematically. The expansion of our knowledge must be paced in line with current knowledge and we cannot take too many leaps into the unknown at the same time.

Nevertheless, we are at a stage today where it is now possible to make such a leap.

It is advisable to make a sensible, logical guess which will not change the predictions or outcomes from proven theory. I have stuck rigidly to accepted science and have only felt the need to deviate where a current belief is a matter of opinion rather than a matter of mathematical imperative or experimentally proven fact. In this regard, I have remained scientifically rigorous and the alternative views expressed here retain their scientific validity whilst offering a different perspective on certain key issues.

I am a Mechanical Engineer by education, not a physicist, although anyone who has studied engineering will know that an Engineer is something of a "pseudo physicist" with a deep and detailed understanding of all things Newtonian and of mathematics in general. Relativity theory is not too far removed from the academics of engineering and I recall my mathematics lecturer at Portsmouth in 1971, taking us through the mathematics of Einstein's field equations at the class's request (it was not actually on the syllabus). The math of GR is gruelling to say the least and indicative of the genius that was Einstein. I just about kept up with it at the time but there are some very good presentations on the internet these days which serve as a refresher. I recommend "Einstein's Field Equations for Beginners" for those readers who have mathematical leanings.

So, while I understand the mathematics of relativity theory for both Special Relativity (SR) and General Relativity (GR), I am at a disadvantage when it comes to publishing the scientific papers needed to present these innovative ideas. This book covers a multitude of topics that are challenging, controversial and in certain instances conflict with mainstream thinking, although not actually with mainstream theory. Rather than attempt to address each point with the individual papers needed to cover the subject matter or take the time-consuming process of using the few scientific publishing channels open to non-physicists, it is clearly more sensible to offer all the elements together in one detailed publication – where they can be examined, considered and judged as a cohesive whole.

In this day and age, practically everyone who is really interested in science has their own pet theory, making it difficult to stand out from the crowd. Such a plethora of bizarre theories exist that it is hard to identify credible new ideas.

To gain acknowledgement from anyone of scientific standing, is more of a challenge today than at any time in the past but, I hope that at least one scientist of repute reading The Binary Universe Theory (B.U.T.) with an open mind, will grasp the ideas presented and will "see the light". Perhaps then, the B.U.T. will stand a chance of being tabled alongside the other candidates which currently aim for unification.

My purpose, in developing the theory and in publishing the work, is to get the ideas on the table. Whatever happens after that, I will leave to providence, and to experimental testing. I will have done all I can.

It may be long after I am dead and gone before the merits of the theory are acknowledged, but the timing, although important to me as a human being, is of a secondary concern to me as an Engineer/Scientist. My motivation has more to do with pointing out truth and with showing the way, than with pursuing fame and glory for myself in my lifetime.

As I write these words, I am sixty-nine years old and when the B.U.T. does eventually break, as I believe it surely will, it may well be presented as someone else's theory, by some blue-eyed boy of the then scientific community.

In 2017, physicists Kip Thorne and Jim Al-Khalili presented, without attribution, one of my ideas published in 2011 (Gravity *is* time)[1], so I can still dream of getting at least a mention in the history books and staking my claim in the ground of human scientific progress.

In the meantime, my rewards have come from the joy of discovery, the acquisition of knowledge, (both old and new), and the sense of achievement from creating this work and bringing it to print. It is difficult not to appear grandiose and egotistical, claiming to have identified the "Theory of Everything" (ToE), whilst the geniuses in the scientific community still struggle to

---

[1] From the 2017 BBC4 documentary "Gravity and Me". I first published this idea in my book "Time Dilation, the reality" in 2011, (ISBN 9780956800206 – page 137 rule number 6.4).

find a breakthrough after more than a century. I believe that despite their undoubted intellectual, mathematical and scientific skills, mainstream scientists have had their noses too close to the grindstone and have simply failed to stand back and look critically at the possible flaws in certain of their opinions and beliefs.

Scientists have no monopoly on intelligence, deductive reasoning, or even mathematical prowess. There are others outside of mainstream physics who do have such abilities and enough knowledge of science to make a meaningful contribution. Certainly, non-scientists have no monopoly on stupidity either and everyone is fallible, even, dare I say it, so are some physicists, at least some of the time. You the reader will be the judge of my credibility.

Some may find my attitude towards the scientific mainstream, provocative, disrespectful or even downright offensive, but these ideas will not gain attention by my being "Mr. Nice Guy". So, please forgive me if I ruffle a few feathers along the way. I intend to do whatever is necessary for the B. U. T. to gain the attention and scrutiny it deserves.

A "fresh" pair of eyes can sometimes reveal mistaken beliefs and unsubstantiated opinions hidden within mainstream scientific theory and there is one such mistaken belief which, on its own, has ultimately led me to all the new ideas presented here.

Engineers are typically presented with a new set of problems on every new project they work on. That is the nature of engineering, so we are perpetual learners. We learn "on the run" then take that new knowledge onwards, applying it through to project completion. We are problem-solvers, compelled to make progress, driven by economic and time pressures. We make things happen and, of necessity, must always educate ourselves technically on the issues at hand to ensure safety, functionality and value. Clearly then, engineers are naturally averse to getting bogged down with pointless philosophical debates and are always keen to take the obvious solution forward, whilst continually reviewing its validity.

So, I feel I am well placed to help sort out the current state of confusion amongst scientists in relativity and quantum physics and to help show the way forward. From my deliberations, the B.U.T. emerges as the logical conclusion from the outstanding issues in today's science and the theory resolves many unanswered questions. When a new idea matches accepted science so well, in addition to providing explanations for many unsolved problems, then there is a very strong case for its consideration as a contender and potentially for its acceptance.

I have had no call on experimental facilities, nor any professional mathematicians to check my work nor any peers to rely upon giving me their opinions and guidance. I have alone and independently identified a new understanding of time that has been under our noses for a hundred years or more and so I do not anticipate any thanks or appreciation from the scientific community for pointing out their shortcomings. Say what they will about me, I am resigned to any critical onslaught or even character assassination that might come my way. I simply cannot keep this to myself, it is too important. I regard myself as fortunate in having the abilities to make this contribution to science and I am compelled to pass this on for the benefit of all, irrespective of any personal consequences.

I respectfully remind scientists that they should not believe in magic, miracles or the occult, and that *everything* in our universe is a physical thing, process or phenomenon. There is no evidence at all of anything beyond the physical and such unfounded ideas require unquestioning faith and unflinching belief without reason. There can be no exceptions to this physicality of everything, but we have lost sight of this when it comes to the passage of time. We have allowed ourselves to get bogged down in philosophical ruminations about the nature of time and we are thus held back by irrelevant and sometimes anthropomorphic obstacles. This is anathema to an Engineer's mind of course, to someone driven to make progress and I have put such irrelevancies aside in my development of the B.U.T.

The book is written and presented in the same sequencing as my learning, my questioning of our understanding, my logical deductions and mathematical workings, so it will take you on the same journey of discovery which ultimately arrives at our binary universe.

Read on and you will see how, by asking simple questions and applying strict causality for the answers, I have been *forced* to deduce a binary universe. You will see *why* the ultimate result from all my deliberations on the subject is the unified field which Einstein failed to define mathematically, yet which was right under his nose. This unified field defines the "Theory of Everything". It is the "fundamental" of our universe from which everything evolves and within which everything exists. It is the physical field of energy that we experience as TIME!

1. Why rock the boat?

Before we get into the details, a general overview will help to understand why it is now most timely to stand back and take stock. There is one, underlying idea that opens up this new path of discovery and leads us towards the Theory of Everything and that is a clear understanding of the physical nature of time, itself. Over the centuries, philosophers have waxed lyrical on the subject of time and there is still much debate over exactly what it is. To me, as an engineer (as should be the case for every physicist), time *must* be a purely physical phenomenon or process. Everything else is physical, so why should the phenomenon of time be any different? As far as we know, there are no exceptions to the purely physical reality of every single action in our universe and there is no reason to suspect otherwise. If we deviate from this, then we enter the realms of religion and psychic phenomenon, of magic, miracles and the occult and that is not science. There remains much confusion about the nature of time. Ideas range from it being just an illusion, purely a facet of our personal perceptions whilst others admit it must have at least some degree of independent, objective reality. Certainly, no one really understands what time is and this ignorance is widely acknowledged. Kant's Block Universe simply does not make sense and many scientists have been misled into thinking that the past, present and future all co-exist together somehow.

> *"Now he has departed from this strange world a little ahead of me. That means nothing. People like us, who believe in physics, know that the distinction between past, present, and future is only an illusion, albeit a stubbornly persistent one."*
>
> *Albert Einstein*

These words are frequently, perhaps unscrupulously used by proponents of the block universe idea to support their outrageous claims. Einstein wrote them in a letter to give comfort to the grieving relative of a fellow physicist who had just passed away. They are about physicists and not about physics. They are not a declaration about any scientific understanding, but they do nonetheless have a certain scientific validity. His statement, I am sure, was an oblique reference to the inertial and gravitational time dilations in Special and General Relativity and the consequential loss of simultaneity between different frames of reference but not, as some people would have us believe, the notion of "all time existing, all of the time".

The past has gone. It is history. It is no longer available. Admittedly, since everyone proceeds into the future at different rates depending on speed and position, then their timings of events might not exactly align with each other and this is what is described by Special Relativity as the loss of simultaneity. This does not mean that the past, everyone's past, remains available to all. That idea is a corruption of logic. It simply means that if you were to stop the whole universe in its tracks instantly, then everyone would be at a different point of progress in time. Their durations and timings of events from the beginning, or from some point at which they *were* simultaneous, are now different.

The theory of Special Relativity (SR), does show that the faster you travel, the slower your time passes (Inertial Time Dilation) and this is mathematically proven, as well as being applied practically in systems like the GPS. General Relativity (GR) also tells us that the rate of time varies with elevation in a gravitational field (gravitational time dilation). We will go into the details of this later but, for the moment, we shall accept these phenomena as a given.

In my years as a professional Engineer, I spent some time travelling in Europe for various clients and on one project I had occasion to fly from Brussels to Alicante every weekend. During the many hours spent waiting in the airport in Brussels, I noticed the horizontal conveyors (or travelators) which took you effortlessly along the airport, to the many departure gates some distance down the concourse. In some places there are more than one, parallel to each other, travelling in the same direction. I imagined people standing on these travelators, each one moving at a slightly different speed and then, when their common power supply is suddenly interrupted, they all break down at the same moment. Clearly, the people on each travelator (frame of reference), have made different amounts of progress along the airport.

A paper glider that was launched forward by a child at the exact moment the people first entered their respective travelators, has now crashed on the stationary airport floor. When the power supply cuts out (coincidentally) at the exact moment of the glider's crash, the time (linear progress) for each person will be different. The crash will not be regarded as simultaneous between the occupants of the different travelators (different frames of reference). Different travelators will each have different periods of elapsed time for any event and events observed in a different frame will seem to occur at different times on the clocks in each frame. The time for any event will depend on which frame or travelator you were on. This is all that is meant by the loss of simultaneity in Special Relativity. It is nothing magical, we simply have to view the whole thing independently, detached from the motion of the travelators, from "outside of time".

You can never go backwards in time. It is done. The future has yet to happen and remains just a host of possibilities without reality, until just the one possibility actually does happen. That leaves us with the present. The present is like the jagged, glowing edges of a piece of smouldering paper with the past burnt away and the intact paper, the possibilities for the future, yet to be consumed. Our universe exists only within the thin red line of the glowing embers, the fleeting present.

The present is the *only* reality even though one man's existence has made a different amount of temporal progress than another's in order to reach the same present. Standing back and observing the two from a timeless perspective shows them both frozen at different stages of development rather than them being jumbled up with time, and with their pasts somehow still available should we wish to call upon them.

If you had been on one travelator and were to cross over onto a faster travelator (frame of reference), and meet the person there, he would have made more progress into his present than you had into yours, but that does not mean you are now in his past or that he is in yours. From the moment of your arrival onto his travelator, you will both be in a common present, it is just that he has experienced more time than you in order to get there. His duration since he originally stepped onto his travelator contains more events than yours. A popular example of this is from the film "Interstellar" where the hero, has been close to (or inside) a black hole and so his time has been slowed down compared to the outside universe. When he finally meets his daughter again, she is an old woman because her time has been proceeding at a more normal, but much faster pace. Yet here too, when they meet, they are both in the same present. It is just that their experiences of time, of duration, have been different up to that moment. She has not entered his past and nor has he entered hers. Neither of their pasts are available to the other. Indeed, he has "travelled into her future" simply by slowing down his rate of existence and letting the outside universe "overtake" him in the temporal sense. This results in her experiencing more time, more change, more events, and therefore more aging than him.

So, time is not about some abstract, ethereal "thing" which remains independent from everyone's reality, it is more personal than that. Time is about change. Local time is about local changes. Time in different locations or at different speeds, will be at different rates of change, just as Einstein has shown us. The daughter in the film has simply undergone more changes than our hero (in the same universal "period"). The ultimate conclusion is that there *is* a universal moment we might call the "Now", the moment, any moment, when we might choose to stop the universe. The loss of simultaneity between frames of reference at any of these moments, (should we choose to stop the universe), is not indicative of some continuous flow of time that we all tap into to the same extent. It is indicative of the number of changes being different in different frames'.

The block universe then, is unrealistic and no one can sensibly infer it from Special or General Relativity. The block universe idea and relativity theory are incompatible, they are mutually exclusive and it is clear to me which one we must believe.

Yet the theory of General Relativity treats time as just another dimension, no different to any of the three physical dimensions that we are all familiar with. Indeed, my travelator analogy treats it the same way, but the obvious differences between time and the physical dimensions have been ignored, shoved under the carpet. No one, until now, has been able to put time in its proper place, to identify exactly what it is in real, physical terms.

Presently, gravitation, as described by General Relativity (GR) does not quite "fit" with Quantum Theory and the book shows how my new understanding of the nature of time helps to reconcile the conflicts between the two. The pursuit of a quantum theory of gravity that fits all scales from the Planck scale to the cosmic scale is ongoing, but there are difficulties in constructing such a theory from the ground up based on a combination of Quantum Mechanics (QM) and General Relativity (GR). Attempts to combine these two theories have led to dubious results and we are at an impasse. Doubt is creeping in about the validity of General Relativity, but it is unlikely that we will identify any errors in our understanding of it purely from mathematical analyses since the "errors" are in the philosophy, not the mathematics. The math of GR works perfectly well, indeed so does the math of quantum mechanics, it is just that they each apply in different temporal domains.

We may never succeed in developing a "Theory of Everything" if we do not attack this problem from both ends of the scale. This book makes the attempt, firstly from the viewpoint of the macroscopic (macro) scale and finally by analysing the nature of space and time at the Planck scale. Employing Logic, we derive a radical, yet intuitive, view of reality itself which is both novel and surprising and yet there is no conflict with accepted science. The theory even explains (or at least clarifies) many present-day conundrums in physics and so, I argue, it must have merit.

In order to fully understand relativity, we have to be clear about what *reality* is and where and when it applies. Traditional Relativity theory confuses relative observational effects with real events and seems unable to distinguish between the two. I am sceptical about the accepted view in relativity which considers all things relative to override reality and am motivated to clarify the resulting confusion over reality itself. Since the advent of relativity theory, realism has become something of a "dirty" word and realists have been given a back seat in scientific thinking. Herein I attempt to redress the balance between reality and relativity whilst respecting both viewpoints. The book identifies where the realities are and where we have been misled by relative observations that are "unreal".

Scientists have no problem in accepting that an image viewed through an optical lens will show a distortion of the real object whilst they know the object itself remains undistorted behind the lens. However, they seem to find it impossible to understand that reality, when viewed through the lens of time, will also show a similar, if only an analogous distortion, again with the reality within its own frame of reference remaining unchanged. (I refer to inertial length contraction in Special Relativity and the fact that proper[2] lengths never change).

Today the emphasis is on particle physics and this is indeed an important field, but we have focussed on this before refining our understanding of the macro world. We dream of the quantum foam, of the intrinsic "energy" of the vacuum, of ten dimensions and of field and particle theories which attempt to explain everything, but we do so without having resolved the discrepancies within the more classical areas of science. Here, I show how our understanding of both the classical and quantum worlds, changes subtly when we resolve these discrepancies.

Despite the progress made by Special and General Relativity, there are still gaps, inconsistencies and unresolved conflicts in our understanding of gravitation and we explain them away with suspect arguments, content with the mathematical rules that we know work in practice. We are *informed* that gravitation in "weak fields" approximates to Newtonian gravitation but that gravitation in "strong fields" requires the full theory of General Relativity to make accurate predictions. In GR, the curvature of space becomes significant in strong fields and this is added to the effects of the Newtonian curvature of time which applies to all fields, weak and strong.

The curvature of space is accepted without question because the behaviour of particles deviates increasingly from Newtonian predictions the stronger the gravitational field. Since 1915 scientists have come to believe this curvature is a real, physical attribute of space, but if you press them, they begrudgingly admit this curvature is merely an abstract, mathematical attribute.

They flit between the idea of real curvature and mathematical abstraction, never quite knowing how to resolve the discrepancy

---

[2] Proper lengths - In Special Relativity, lengths appear to contract in the direction of motion with increasing speed as viewed in the moving or accelerated frame of reference, but the length within the moving frame, the "proper" length never changes.

and so they give up on the issue. This is where we still are today, over a century on. There have always been dissenters regarding the idea of the curvature of space and I have found myself among them, for the same reasons.

> *"I hold that space cannot be curved, for the simple reason that it can have no properties. It might as well be said that God has properties. He has not, but only attributes and these are of our own making. Of properties we can only speak when dealing with matter filling the space. To say that in the presence of large bodies space becomes curved is equivalent to stating that something can act upon nothing. I, for one, refuse to subscribe to such a view."*

> *Nikola Tesla*

As you will see later, both Tesla and Einstein are right and the B.U.T. brings together these two superficially opposing views. It is not my intention to criticise General Relativity (GR), in any way but merely to explain it with a complete causality. As you read on you will find my ideas AGREE with GR and I unreservedly support the theory. What I am saying is that you cannot have the curvature of space without a direct cause for the effect and I will demonstrate this causality later. GR is the best theory of gravitation we have and it does give correct predictions. So far, it has passed every test applied to it. Clearly, it works and its predictions are accurate. Nevertheless, it cannot be justified at the fundamental level. It cannot be explained by logic, deductive reasoning, common sense or insight. There is no clear causality. It is only the mathematics which fully "describes" the behaviour. But the math is based upon a flawed idea. It relies on the abstract notion that space itself is "curved" by the presence of mass, yet there is no suggested mechanism or cause for this effect even if it were possible to physically curve what is, essentially, nothing at the macro scale.

I am not denying that a beam of light follows a curved path in a strong gravitational field. I am questioning how space can actually curve as if it were a real physical attribute. Curved trajectories or geodesics are an *effect* within space-time, not a cause, not a property of space. Later, I will show how GR can indeed be explained not just in terms of Einstein's abstract geometry but also in terms of a logical understanding of the real, physical and temporal world and with a known cause and effect.

Until recently, cosmologists and quantum physicists have all viewed space as flat in their respective fields and it was only relativity that had allowed it to be curved. Presently, there are attempts to reconcile GR with Quantum theory including Loop Quantum Gravity and String Theory but so far, without verifiable success. There is a persisting disagreement between various scientific factions over the validity of string theory, but some articles present string theory as a given. It is not. It is more that string theory has been around for so long that it gives the *illusion* of its having been proven and accepted.

The justification for this space curvature is simply that, mathematically speaking, there is no reason why lines must be straight or planes flat and no reason to believe three-dimensional space can only be accurately described using straight lines and flat planes. Einstein applied a whole new geometry based on the mathematics of Gauss and Riemann involving an abstract, mathematical curved space and this non-Euclidian geometry is the basis for the mathematics of General Relativity. In other words, he starts off with the idea that mass *somehow* causes the curvature of "nothing" and develops his theory from that basis.

His theory starts from this hunch, a guess, and I concede that viewing space and time in this geometric way *is* an effective and efficient way of developing a theory of gravitation. This does not mean that space curvature is a real, physical phenomenon, but only that this "curvature" approach will get you to the right answers. General Relativity actually admits that space curvature is abstract (i.e. not real) so, I think that Albert and I might have agreed on this, but the modern mainstream has lost sight of this unresolved conflict and has shoved yet another critical conundrum under the carpet.

GR is a *geometric* interpretation of space-time which does not necessarily reflect reality. It certainly does not provide a stringent causality. The math gives the right answers, at least on the macro scale, but the theory does not explain the fundamental cause and effect of gravitation. It does, however, provide a way of accurately predicting behaviour using rigorous mathematics.

Despite the success of GR, we have to face the fact that abstract ideas cannot move physical objects and we seem to have forgotten that only a real physical cause can have a real physical effect. Unlike some modern-day physicists, I and the likes of Nicola Tesla, refuse to give up on causality quite so easily.

The mathematics of GR is analogous to a four-dimensional draughtsman's scale[3] which has all the rules of physics built into the geometry. Since it is constrained, both by the laws of nature and by mathematical rigor, it cannot fail to give correct predictions. The beauty of mathematics lies in the fact that while it will always describe and even define real behaviour, like the draughtsman's scale, some of the intermediate workings may have no physical meaning and they are simply a clever, reliable technique for getting to the right answers rather than for describing reality. General Relativity avoids the necessity for rigorous, physical causality and instead uses the magnificent "trick" of non-Euclidian geometry for producing solutions to space-time analyses.

I have one more question arising from GR. We know that in nature there is invariably only *one* fundamental cause for any fundamental effect and I assert that gravitation can be no exception to this rule.

This problem has been skilfully avoided by viewing space and time as one entity, "space-time", but this book shows that we should view space and time separately. Certainly, you cannot have space without time, but they are fundamentally different, despite being inextricably linked, and I will demonstrate how time is *the* fundamental and that space actually emerges only when time starts to pass.

My hunch, or guess, before arriving at the B.U.T, was that there can be only one fundamental cause for the acceleration of moving objects in both strong and weak gravitational fields. This cause cannot be the curvature of time for weak fields and increasingly, the distortion or curvature of space, the stronger the field. It can

---

[3] Draughtsman's Scale – A method to divide a line of unknown length into a number of equal lengths. A "Trade School" method using a ruler and a square.

only ever be the curvature of time and its associated effects as the field strength increases.

Time can be shown to be curved in real terms, e.g. the time rate varies with elevation in a gravitational field, but no one has yet been able to demonstrate *how* space can "curve", only that it *does*, in an abstract mathematical context.

Logically, the distortion effects, when events in strong fields are observed from elsewhere in the field, can only be due to the differential time rates and how they vary between the observer and the event. Nevertheless, it is *as if* space is physically distorted as well as time being *curved*, so this is why General Relativity works. It may as well *be* that space is curved and it is therefore easy to fall into the trap of believing this to be real and to just accept it without any causality. Physicists today appear to accept this notion of physical curvature without question and to dismiss any ideas to the contrary, entrapping themselves in their own cage of fixed and rigidly-prejudiced thinking. I will demonstrate that it is the *rate* at which time curves over distance which results in the effective curvature of space. General Relativity is correct. It will always give the right answers, but it does not give a real cause and effect. I will demonstrate that later.

String Theory[4], perhaps the leading contender in our attempts to explain our universe at a more fundamental level, inherits this imaginary idea of space curvature from the incorporation of GR into the math. It thus becomes partly abstract, as it retains this unreal phenomenon, but nevertheless remains mathematically rigorous like GR itself. Yet it is not possible to describe reality using an abstract notion, however rigorous the mathematical treatment, but only by using real phenomena and strictly adhering to physical and temporal cause and effect.

Scientists must always look reality in the face and not be beguiled by the continued experimental verification of GR into blind acceptance, or be complacent due to the apparent lack of a need to question, nor should they be afraid to challenge such a revered theory and its highly regarded protagonists.

Today, the cracks are starting to appear in the form of

---

[4] String Theory – A mathematical theory envisioning space time as vibrating strings at the very small scale.

anomalies, such as the Galactic Rotation Anomaly[5], the Pioneer Anomaly[6] and the FlyBy Anomaly[7], none of which are predicted by accepted science, so they are not understood by present day scientists. The "final" report from NASA on the Pioneer anomaly suggests that the anomaly has been "talked away" as an irritatingly persistent discrepancy between current scientific beliefs and actual observation. In my opinion, having read the report, it still remains an anomaly.

We have invented "Cold Dark Matter" (CDM) and "Dark Energy" to explain certain observations and, although only speculative explanations are put forward for these phenomena, they are now considered as a "given". Science has again closed its mind to any different interpretations of these effects.

Many modern-day scientists expect mathematics alone to break the new ground of our understanding and they have thrown in the towel regarding insight, visualisation and deductive reasoning. They have lost faith in the ability of the human mind to grasp the realities of our universe and have decided to believe we lack the intellect to do so. This defeatist attitude is now a major obstacle to scientific progress and science places too much emphasis on mathematics, dismissing deductive logic and a visualisation of the physical world. In fact, if you present to the "mainstreamer" a logical, deductive argument which disagrees with the mainstream view in any way, you are likely to be accused of "Philosophical hand waving" and your point, valid or not, is immediately rejected. We are at a stage, today, where there is no longer any room for debate and that is a dangerous place to be, for our scientific progress. We are in the age of the "Mathematical Physicist" but mathematics and physics are different and, in some cases, they appear to describe different realities depending on the interpretation of the math.

We rely on classical mechanics and the mathematics of Newton and Einstein to make successful predictions and to engineer solutions to problems, without really knowing what we are doing at the fundamental level. Surprisingly, we are still being told that

---

[5] Galactic Rotation Anomaly – The rotation speeds of stars in spiral galaxies are faster than Newton's laws predict.
[6] Pioneer Anomaly – Unexplained *blue shift* of signals from the pioneer probes.
[7] FlyBy Anomaly – Unexplained red and *blue shift* of signals from various probes at perigee when flying past the Earth.

gravity is some mysterious *force*, which is yet to be fully understood and reconciled with the other forces in nature and we are discouraged from questioning this "belief". This is despite the fact that the gravitational "force" breaks the causality principle[8], a fundamental law of nature. I will show, independently of GR, that gravitation is not due to a force of any description, even though it may be treated as one mathematically in some circumstances.

Scientific thinking seems to have been misdirected away from a logical approach which uses deductive reasoning and sensible assumptions to gain a clear understanding of the macro universe[9]. Some have even said that they do not expect there to be significant advances via a classical, logical analysis from now on. I believe this is a huge mistake, as I do not think we have examined all possibilities from such an analytical approach using currently available knowledge.

Gravity and inertia are *the* major impediments to space exploration and it is surprising to me that we have neglected the pursuit of a better understanding of these phenomena for so long, despite our total reliance upon their rules of behaviour. Put simply, we do not know what a clearer, more realistic view of gravity might bring to our understanding of the universe in general and to our understanding of the quantum universe. Attempts by quantum physicists to find a quantum cause for gravitation have so far not yielded a solution but this book provides a different way of looking at space-time which will have a major impact on the way we might view the quantum world.

The familiar rules of behaviour which operate at macro scales appear less relevant at particle scales and even less so toward the Planck[10] scale. This clearly inhibits progress using common sense analysis, at least for the moment. However, this does not mean a logical approach will not bear fruit in the pursuit of a greater understanding of the "macro" world. It is my intention to provide a more accurate, simpler and realistic view of relativity and I

---

[8] Causality Principle – For every effect, there must be a proximate, antecedent cause.
[9] Macro universe – The size of the atom and larger.
[10] Planck Scale – First identified by Max Von Planck (1858 to 1947). The fundamental, natural units of space and time and are labelled – Planck length and Planck time. There is also a Planck mass together with all other physical quantities.

respectfully suggest that physicists in quantum mechanics (QM) should review and reassess the relationship between QM and the Special and General theories.

The book focuses firstly on the macro scale and takes a logical, deductive approach to investigating the nature of space-time and gravity. The basic philosophy of Special and General Relativity is analysed and straightforward "High School" mathematics clearly demonstrates that certain current beliefs are indeed mistaken, and that the mainstream mathematics of both theories does allow this new interpretation.

Later, we focus down to the quantum scale and an in-depth analysis of the nature of time itself is presented. Ultimately, this leads us to the inevitable conclusion that we live in a Binary Universe where there is a real, negative mirror of our positive binary half.

I offer here a clarification of Special and General Relativity which identifies the cause and effect for inertial time dilation, gravitational time dilation and the effective curvature of space in strong fields.

Finally, as a result of these deliberations, I present the inevitable conclusions about the nature of time and of the universe itself from the Planck scale upwards, a clear basis for the "Theory of Everything".

The book expands our understanding of gravitation beyond Newton's fictitious but convenient "force" acting at a distance, which is useful in practical terms, but which breaks the laws of physics. Newton's force is incapable of being satisfactorily explained since the notion of "action at a distance" breaks the principle of causality. In fact, the curvature of time and space from General Relativity has already replaced Newton's force acting at a distance with a field effect, yet the mainstream still hangs on to the idea of some "force" of gravity and this will remain a conflict until the nature of this field is better understood. Here, I identify that our gravitational field is not a physical entity but is a fiction, created by us to explain the behaviour within it. I propose the only viable alternative which explains the abstract curvature of space in real, physical terms. The proposals made here are based on physical and temporal cause and effect and on directly detectable phenomena, not just on what we might *infer* from observed behaviour. Inference is a risky process after all.

Mainstream science acknowledges that gravity goes hand in hand with time dilation[11] or *red shift*[12], but suggests no cause and effect between the two. It is accepted, blindly, that time dilation is present in any gravitational field and that you cannot have one without the other, yet the causality is neglected. One either has to find a physical cause for both of these effects (gravity and time dilation) or, better still, define the causality that links them. This is where these proposals differ subtly, yet importantly, from current beliefs and this long-neglected issue is finally resolved here. The causes for *how* the presence of mass slows down time and for *how* the rapidly changing time rate over distance effectively curves space, are both demonstrated.

Because these hypotheses have evolved from accepted relativity, because of the sound deductive reasoning used, the elimination of conflicts and discrepancies in current science, the mathematical proofs, the accurate prediction of Newtonian gravitational acceleration, together with full compliance with the causality principle, certain of the hypotheses are proposed as addenda to the theories of Special and General Relativity. The mathematics used for these theories works perfectly and so there has been no pressing need to change them. There is however a pressing need for us to change our *ideas* about them.

> *"The important thing in science is not so much to obtain new facts, as to discover new ways of thinking about them."*
>
> Sir William Lawrence Bragg (1890 – 1971)

Quantum field theory[13] and, in particular, the Standard Model[14], is recognised as incomplete and does not have an explanation for the "force" of gravity. The Graviton[15] particle is proposed as the "carrier" of the gravitational force.

---

[11] Time dilation – The slowing down of time. The "curvature" or change of the rate of passage of time with position.
[12] *Red shift* – A change in frequency towards the longer wave length, in the direction towards the red end of the visible spectrum.
[13] Quantum field theory – The basis of particle physics
[14] Standard model – The model of elementary particles and their interactions.
[15] Graviton – theoretical subatomic particle proposed as the purveyor of the gravitational force.

The reason for this postulate, is simply that the other forces in quantum field theory are believed to be caused by the interactions of subatomic particles and it is therefore *assumed* that gravitation will be likewise produced. More recently, quantum field theory does identify the field as the active entity and not the particle associated with it. The field does the work, but the particle is simply something that is ejected from the field, if you impact it with a particular energy. The graviton, from the field of gravity, has not yet been observed (and I believe it never will be if the proposed theory survives the tests that I hope will ensue as a result of this publication). Indeed, modern scientific papers do now refer to *virtual* gravitons, half admitting they do not really exist and are merely imaginary, mathematical abstractions which are necessary to make the mathematics of quantum theory work. The field that causes gravitation is identified here, along with its particle, and I am sure this discovery will be a great surprise to most physicists. Certainly, I have been unable to discover any published papers, as yet, on the subject.

The new theory shows that gravitational attraction is caused, not by a force, but by a field effect which is *equivalent* to a force that *would* obey Newton's laws if, indeed, it existed. Einstein did refer to an "equivalent force" and General Relativity does provide this field effect, but the field in GR is one of time *and* space curvature. The gravitational field in GR is therefore partly abstract in explaining the deviations from Newtonian laws in strong fields. The thing is, you cannot get a particle from impacting an imaginary field and our inability to find the graviton confirms that the gravitational field is unreal... it is our creation. This book identifies the completely *real* field which is the cause of gravitation in both weak and strong "gravitational fields".

Much has been written on the subject of a "Unified theory of everything", but so far, no one has come up with a sensible proposal for how gravity works or how it interacts with the quantum world. Gravity is the only "force" that presently cannot be unified with the other forces in nature, the "electroweak" and "strong" forces and our incomplete understanding of gravity is therefore a major stumbling block in the pursuit of a unified theory. Herein, we identify the fundamental building block of all matter, indeed the fundamental entity of our universe which points the way towards the "Grand Unification".

In my writing, I have tried to stick to the point and avoid unnecessary complexity. I have aimed to explain things for the benefit of the layman, keeping the contents within the realm of understanding by the technically minded, motivated reader who perhaps has more than just a passing interest in physics or cosmology and relativity. Despite mainstream resistance against alternative thinking, especially when it originates from outside the scientific community, these issues demand a better explanation than we are currently offered and there are plenty of reasons to challenge certain, current misconceptions... many reasons to rock the boat.

2. Some historical scientific developments

With the benefit of hindsight, some historical beliefs (ultimately proven false by science) may seem ridiculous to us now, but it is worth taking a close look at a few, to see how they were eventually replaced.

In turn, this should make us more sceptical of some of today's unproven ideas, challenging their validity until they either stand proven, or fall and are replaced with new, better ideas.

These mistaken beliefs prevailed, not because people in the past were any less intelligent or capable than those whose ideas prevail today, but because it is in our nature to try and understand our observations and to try and justify them.

We make progress with our understanding using sound science, experimentation, observation, mathematical proofs, sensible assumptions, logical deduction, empirical methods, and, in the absence of some or all of these, even by creative thinking.

We are all made that way.

When we face the unknown, our imagination tries to "join the dots" and we risk coming up with ideas that may not withstand the rigours of scientific analysis. The positive side to this is that we hopefully come up with *all* the possibilities and then investigate each of them to arrive at the scientific truth, however long that might take.

In the development of the B.U.T. I have controlled my creative thinking (and its associated risks) by ensuring full compliance with existing theory, experimental evidence, sound philosophy, deductive reasoning, verifiable assumptions and have provided where necessary, just a little mathematics.

The Earth at The Centre of The Universe

The geocentric or Ptolemaic view of the universe originated in ancient Greece. It supposed the Earth was at the centre of the universe and all the heavenly bodies circled the Earth. This theory, further developed by Ptolemy[16] relied on several arguments including the following:

- If the Earth did move, then one should be able to observe shifting of the fixed stars due to parallax[17] and this was not evident. The alternative was that the stars were at extreme distances from the Earth and therefore showed negligible parallax, but this idea was dismissed in favour of the static Earth.

- The brightness of Venus is fairly constant and this was interpreted to mean that it remained at the same distance from the Earth all the time. The reality, of course, is that the varying distance is largely compensated for by the waxing and waning crescent which gives the impression of constant brightness and therefore distance.

- Half the stars were observed above the horizon and half below at any one time and this was understood to mean that the Earth must be at the centre of the universe, since if it were significantly displaced the division of visible and invisible stars would not be equal. The assumption was that all the stars were at some modest distance from the centre of the universe, the Earth.

To try and make the Ptolemaic system match the observed variable brightness of all the other planets, an "orrery" was created, a complicated mechanical system of gears, spindles and rotating spheres, that ensured the necessary variable planetary distances to explain the observed brightness variations. This is an

---

[16] Ptolemy – Claudius Ptolemaeus, a Roman citizen of ancient Egypt.
[17] Parallax – Difference in apparent position of an object viewed along two different lines of sight, measured by the angle between those two lines.

excellent example of how clever and creative some scientists can become in their determination to justify current beliefs. The most eminent scientists of the day were quite capable of constructing complex and perfectly viable theories which matched observations and made accurate predictions but which did not reflect reality. What they could not do was to determine the cause and effect of the differing motions of the planets within their complex machine since this was an insurmountable problem for them, entirely understandable when considering the limited extent of scientific knowledge at the time.

In 1543, Copernicus published his theory that the Earth and other planets revolved around the Sun. In 1610 Galileo Galilee used his telescope to show that Venus had all the phases one would expect in a heliocentric[18] system rather than the Ptolemaic system. Also in the early 1600s Kepler devised his three laws and proposed a heliocentric system with elliptical planetary orbits. In 1687 Newton presented his laws of motion and gravitation thus allowing a plausible heliocentric system. Eventually, in 1838 Friedrich Wilhelm Bessel successfully measured the parallax of star 61 Cygni, finally disproving Ptolemy's assertion that parallax motion did not exist.

Whilst it may have been evident to some, along this tortuous route, that the Ptolemaic system was not the real truth, it nevertheless took the best part of 2,000 years to finally kill it off. Despite the virtues of the proposed alternative heliocentric system the Ptolemaic system hung around until scientifically disproved. It was ultimately the technological advance of more accurate measuring techniques that allowed Bessel to measure the parallax, and so disprove the Ptolemaic theory once and for all. In summary, the transition from the Ptolemaic theory to the heliocentric system involved:

- Explanation of the phases of Venus. (In spite of this contradiction, the Ptolemaic theory persisted)

- The postulated new theory of heliocentricity. (This still did not override the Ptolemaic theory and both theories co-existed in contradiction for a long time)

---

[18] Heliocentric – A system where all the planets revolve around the Sun.

- Newton's laws of gravitation showed how the heliocentric system worked. (This seemed to have been partly ignored and the Ptolemaic system persisted at least in the minds of some)

- The proof of parallax with the stars finally added enough weight to the heliocentric argument and the Ptolemaic system finally collapsed.

This example of scientific progress shows just how difficult it is for any new theory to become accepted over entrenched ideas. A huge weight of evidence is required to overcome what can only be described as a scientific prejudice in favour of the prevailing beliefs of the time. Today the burden of proof still rests, quite rightly, on those who challenge the mainstream, but there seems to be great difficulty in assessing which of two theories is the better, especially if one has been around for a long time and the new one disagrees. Only a clear mathematical or experimental proof will succeed in dislodging such prejudice.

The difficulty with the B.U.T. is that it does not disagree with any of the mathematics or experimental results, nor with any current predictions except one! This prediction is that different frequencies of electromagnetic radiation (EMR) must all travel at very slightly different speeds, just below the speed of light. This one prediction has already been demonstrated, in practice, but physicists are attempting to explain it away – confined by their entrenched beliefs and closing their minds to the real, alternative explanation. The surprising thing is that the higher frequencies travel the slowest. We shall see why, later.

This "inertia" with scientific progress is not always a bad thing and it does protect accepted science from corruption by "Crackpot" ideas.

New ideas must always go through the necessary stages from ideas to postulates, to theories, to mathematical proofs, experimental confirmation and eventual acceptance. Then the whole thing starts again when someone comes up with an alternative idea which may seem to contradict, and so it goes on.

The Flat Earth

The idea of a round Earth has been acknowledged in the Western world since 570 BC when Pythagoras first postulated that the Earth must be a sphere. Since then the idea of a round Earth has been alive in the scientific community (and not nearly as controversial as most of us are led to believe today).

This concept of a spherical Earth was in dispute for the best part of 300 years and was not resolved until the third century BC when Hellenistic[19] astronomy established the round Earth beyond reasonable doubt.

By 330 BC Aristotle had accepted the idea of a round Earth and from this time onwards, this understanding spread beyond the Hellenistic world.

There has been a historical misconception in the West that the prevailing cosmological view at the time of Columbus was that the Earth was flat, not spherical and this misunderstanding has been referred to as "The myth of the flat Earth".

This misconception flourished mainly between 1870 and 1920 and was connected with the ideological struggles over evolution.

> *"With extraordinarily few exceptions, no educated person in the history of Western Civilisation from the third century BC onwards believed that the Earth was flat."*
>
> *Jeffrey Burton Russell*

The final practical proof of the spherical shape of the Earth was not demonstrated until the voyage of Ferdinand Magellan who circumnavigated the globe in 1519-1521.

This short overview serves to show just how long it can take between a postulate being formed and the experimental proof of its validity, in this case about 2,090 years.

---

[19] Hellenistic astronomy – Astronomy of the Hellenistic world, the world as conquered by Alexander, the main language of which was Greek.

The Aether

The idea of the Luminiferous Aether[20] was created in order to visualise the propagation of light through an otherwise empty void or vacuum.

The logic was that, just as ripples in a pond needed the medium of water to propagate through, then the transmission of light, being of a wave form analogous to ripples, also needed a medium to transmit through.

The Aether was a concept invented by Christiaan Huygens in the seventeenth century and it was Huygens who postulated that light was of a wave nature. It was simply inconceivable to the scientific thinking of the day that anything of a wave nature could be transmitted through nothing. In fact, the wave form of light is still an issue today in the absence of the Aether and we will see later just how and why light takes on its wave form as it radiates through the vacuum.

Scientific progress, away from the Aether, was complex and protracted, involving many scientists and experiments. Initially, Newton regarded light as being of a particle nature since there was then no issue in visualising light being transmitted through a vacuum or with it being reflected. Newton had rejected light as waves in a medium because:

> *"Such a medium would have to extend everywhere in space, and would thereby "disturb and retard the motions of those great bodies" (the planets and comets) and thus as it [light's medium] is of no use, and hinders the operation of nature, and makes her languish, so there is no evidence for its existence, and therefore it ought to be rejected."*

> *Isaac Newton*

In 1720 James Bradley carried out an experiment to prove stellar parallax, but could detect none. However, he did detect stellar

---

[20] Luminiferous Aether, or Aether, or Ether - the theoretical medium permeating space through which everything travels. Now superseded by General Relativity.

aberration[21] with the stars changing their apparent positions depending on the position of the Earth on its orbit around the Sun. This effect is dependent upon speed rather than distance. So, by adding the vectors of the known orbital speed of the Earth to the velocity of the light vertically down, and knowing the angle of aberration, this enabled him to estimate the speed of light.

In the Aether based theory of light, this effect would have been very difficult to explain since it would have required the Aether to be stationary even as the Earth moved through it.

A century later, Young and Fresnel pointed out that light could be a transverse wave rather than a longitudinal one and Newton's particle theory of light was abandoned. Scientists still hung on to the idea that light waves needed a medium to propagate through, which required Huygens' Aether gas permeating all space.

The next investigations and debates involved complex issues such as the physical and electrical properties of the Aether and how it could be envisaged in order not to challenge known science.

Again, it is worthy of note how determined scientists can become, when a theory does not match reality and is in its death throws. The Aether had to be fluid, without mass, have zero viscosity, be completely transparent, non-dispersive, incompressible and completely continuous at a very small scale, and yet it was still believed by the scientific community of the day, to exist.

To my mind, this situation is not dissimilar to the scientific community's current belief in dark matter, another miraculous entity with inexplicable, magic properties. As of writing, there are now articles appearing in science magazines questioning the existence of dark matter and serious investigation is now beginning.

It was not until the Special Theory of Relativity in 1905 that a viable alternative was available and the Aether finally fell to Occam's razor. In other words, it all got far too complicated to hang on to, when compared with newer "simpler" theories.

---

[21] Aberration - Apparent motion of celestial objects caused by the orbital motion of the Earth around the Sun and the finite speed of light.

Occam's Razor

Occam's razor is the meta-theoretical principle that generally recommends that a "simpler" hypothesis which makes fewer assumptions and avoids creating new entities is preferred over a more complex, competing hypothesis that gives the same result. In short, the simplest explanation is the most plausible, unless or until evidence is presented to prove it false.

Of course, the definition of "simpler" depends on one's point of view and this changes as knowledge increases. There is much written on the "razor" and still much debate, but the above definition will suffice for the purposes of this book.

Although Occam's razor is not an irrefutable principle of logic, it is, nevertheless, an accepted scientific rule of thumb or guide, so the reader should bear this in mind when comparing proposed new ideas with their mainstream alternatives.

So, What Can We Learn From History?

Even from these brief overviews of historical scientific developments, we can see how difficult it is for a new theory (or modifications to existing theory) to become accepted, to get rid of entrenched ways of thinking and establish a better understanding of the world around us. We see that the requirements for change include – original thought, personal courage, technological progress, new ways of looking at the accepted view, the co-existence of different theories until only one is left (and all the others disproved), mathematical proofs, experimental verification, the accurate interpretation of observations, continual questioning and much tenacity, sometimes over many generations.

We see that prevailing scientific beliefs are not, necessarily, correct but are invariably regarded as such. If such beliefs have existed for long periods and much work relies upon their correctness, then this can give rise to a disproportionate resistance to change. This can occur, for instance, when we have created spurious ideas and entities and have put in a great deal of effort over many years to justify them, to make them fit current theory and observation. Changing from such a situation might cause great disruption, additional work and any challenges to these beliefs are likely to be a source of embarrassment to the scientific establishment.

Perhaps the greatest obstacle to us changing our thinking is when a misconception or unreal idea actually works for all practical purposes.

In General Relativity (GR), the curvature of space coupled with the curvature of time does give correct solutions and so, naturally, there is little motivation to question the theory. Even so, it must be of concern, even to mainstream physicists, that the curvature of the vacuum is an abstract notion and that abstract ideas cannot accelerate physical objects. This effective curvature of space requires a better explanation than simply, "The mathematics tells us so" and I will explain the real, physical causality later.

Today, there is no reason to suppose that we have come to the end of the process of scientific development or that we are approaching a complete understanding of the physical world.

Neither should we assume that our present understandings are completely accurate as there is always room for questioning, even at the most basic level. Understanding this may give us more awareness of current uncertainties and perhaps a little healthy scepticism about certain aspects of today's "accepted" science.

3 Some flaws in present day science

Accepting that the scientific knowledge of any era has always been subject to challenges, we must also accept that current theories ought to be the subject of continual scrutiny today. We will now identify areas where there are conflicts, discrepancies, unsubstantiated assumptions, neglected issues and other aspects which cast doubt over certain opinions and beliefs within accepted theory.

Gravitation

Even today there is no consistent explanation for what gravity is, how it is created or the mechanism by which it operates. Gravity has just been accepted for what it *does* and for the rules it follows simply because we can find no issue with it, at least for the moment.

> *"In terms of the sphere of human activity, we understand enough about these two fields (gravity and magnetism) that we will never have much practical need for better theories than the ones we now have."*
>
> *Dr. "O"*
> *(Name withheld to avoid embarrassment)*

This statement is a prime example of the complacency and arrogance displayed throughout history by a minority of scientists. We just do not know what an expanded knowledge of gravitation and inertia might bring to science and engineering.

For us to be missing potential opportunities any longer than necessary, simply because we feel comfortable with our current understanding is quite unacceptable.

We know the rules that gravity obeys, namely:

> *Gravitational acceleration is equivalent to the application of a force which is proportional to the attracting mass, always in the direction towards the centre of that mass. This force decays with distance from the source according to the inverse square law.*

This knowledge has been sufficient to get us to the Moon, to send probes to the outer planets of the Solar system and to develop a permanent, manned presence in Earth orbit, so "Why," you may well ask, "is there a need to know more?" Well, just as space exploration is the natural pursuit of knowledge and our expansion into the unknown, then so we should take the same approach to expanding our scientific knowledge. This may not be a physical frontier, but we should consider what physical frontiers have been opened up and pushed back by our greater understanding of physics, remembering that space exploration has been one of them.

Whilst historically, necessity has tended to be the "Mother of Invention", it is possible to make scientific progress and to expand our knowledge without waiting for some pressing need, like a world war or an international race to the Moon. In fact, the ancient Greeks, Galileo, Newton, Einstein and many modern-day scientists have all made scientific progress thanks to their innate curiosity and thirst for knowledge rather than to any other driving force - funding requirements having been met of course.

Today, there is still no satisfactory explanation for why objects with different masses, in the same orbit, are both kept in that orbit by gravity. You might think that the heavier object would fly off and settle in a larger orbit, due to its greater tangential momentum. The point is that, in the same orbit, each object has a different energy of motion, due to their different masses, yet gravity seems to ignore this fact and simply keeps them both there as if they were identical.

Imagine if you swing a one-kilogram object around you on a one metre length of string and then do the same with a two kilogram object. You will certainly see that you need to apply twice the force to restrain the two kilogram object at the same speed of rotation, yet gravity seems to be very selective about the "force" it applies to individual objects and we observe that gravitational "free fall" is independent of the mass of the object being attracted. If gravity were a force, then this force would have to be applied proportionate to the free-falling mass and this is physically impossible. There is no piece of string with which to exert these different forces. The challenge is to find out how gravity does this (and not to simply brush the issue under the carpet using suspect argument, or to give up without identifying the reason). Presently, we rely on the Principle of Equivalence[22] without fully understanding the causality. There are mathematical formulae that predict all this observed behaviour, even taking account of relativistic effects, but there is still a lack of understanding and visualisation of the what, how and why.

There is still no satisfactory explanation, other than the mathematical rules of behaviour, for why the classic hammer and feather, when dropped at the same instant (by Apollo 15 astronaut, Cmdr. David Scott), both fell to the Moon's surface with the same acceleration value. As predicted, both objects hit the ground at the same time. You might think that the *force* of gravity would pull the lighter feather more quickly than the heavier hammer, but it is not the case and this is another example of how gravitational free fall is independent of the mass of the body being attracted. This anomaly has been around since Newton and, although it has been addressed mathematically with Einstein's Principle of Equivalence, we do not really know why inertial mass[23] is equivalent to gravitational mass[24]. This equality is put down to one of nature's highly coincidental and convenient rules.

For these reasons, it is clear that gravitation is not due to the application of force, since it accelerates all bodies by the same

---

[22] Principle of Equivalence – states that standing in a gravitational field and experiencing the pull of gravity is equivalent in all physical respects, to being accelerated upwards in space with the same value of acceleration.
[23] Inertial mass – The mass of an object which produces inertial resistance to acceleration.
[24] Gravitational mass – The mass of an object, equivalent to inertial mass, which determines the degree to which it is affected by, or generates, a gravitational field.

degree, irrespective of their masses. If gravity were a force, this would not be the case and, with gravitation expressed as a force, Newton's laws themselves are compromised by the Causality principle[25.]

Einstein developed the Principle of Equivalence between gravitational and inertial mass which demonstrates that this "equal" behaviour is a rule that everything must follow and this is deduced from Newton's Law of how objects behave in a gravitational field. Einstein uses these observations and then applies Newton's Second Law[26] to deduce the Equivalence Principle. However, the Equivalence Principle still does not explain *how* gravity does this, but merely provides the rules of behaviour.

The major issue, with Newton's force, is one of cause and effect. Newton's theory relies on his force "acting at a distance", which breaks the Causality principle in physics. Newton himself was aware of this and these were his own words on this point:

> *"That one body may act upon another at a distance through a vacuum without the mediation of anything else, by and through which their actions and force may be conveyed from one another, is to me so great an absurdity that, I believe, no man who has in philosophic matters a competent faculty of thinking could ever fall into it."*
>
> *Isaac Newton*

He refers to his force acting at a distance and says that anyone with even half a brain should never believe it! Of course, he was right and, today, most of us do not believe in action at a distance. To overcome this serious flaw, science has proposed the "Graviton" particle, which must radiate outwards from a mass and which somehow ends up attracting everything in accordance with Newton's laws. I will leave the credence of this idea to your own instincts. The Graviton has still not been proven to exist.

---

[25] Causality principle – states that for every effect there must be a proximate, antecedent cause.
[26] Newton's second Law – Force equals mass times acceleration, $F = ma$.

We do have two realistic choices when considering the cause of gravitational acceleration. The first suggests force applied at the point of acceleration and the second is that there is some other reason, yet to be explained. Clearly, since Newton's force cannot be applied locally at the point of acceleration (or indeed to varying degrees for different masses), then we must conclude that there is indeed some other reason for the acceleration.

Relativity deals with this problem with a mathematical model of space-time which has curved "World Lines[27]" or "Geodesics[28]" representing the "shortest" or "easiest" paths for an object to follow through space-time, lines of "least action". Because we observe curved paths of motion for any entity, including light, in a gravitational field the idea of curved space-time and of curved space, has become widely accepted. The current popular example of this is gravitational lensing, where the light from distant galaxies is clearly bent two ways around some massive unseen entity between us and the light source. The light appears to us as an arc of light, duplicated in two opposite positions. Clearly the light has followed a curved path, but later in the book we shall see that this can be explained without naively accepting the physical curvature of space.

The mathematics which defines these space-time models is complex, yet nature does not have to be that complicated for it to produce the effects of gravitation. Indeed, present day scientists do now accept that Newtonian gravitation can be regarded as the curvature of *time* and space curvature is not required to explain pure Newtonian gravitation (weak fields). However, relativists do not accept the idea of a universe with a "flat" space-time that might result from this view and the scientific community has chosen to follow the curved space of General Relativity.

The theory proposed here postulates *how* gravity is created by the presence of mass and unlike the mainstream, does not invoke an act of faith, or blindly accept that gravitation is, somehow, always associated with mass without investigating the causality. The theory postulates its cause, then shows *how* gravity works and *why* it obeys the rules that it does. These explanations are achieved without recourse to the complex mathematics of General

---

[27] World line – a curved path through a distorted space-time along which objects naturally follow.
[28] Geodesic – Another name for a World Line.

Relativity, but they do involve a basic understanding of the Special and General theories. The theory requires no new entities and is the simplest and least problematic explanation put forward today. In these respects, at least, it threatens certain aspects of current theory with Occam's razor. Most importantly, the proposals are fully compliant with the Causality principle and generally compliant with the theories of both Newton and Einstein. For the cause and effect of Newtonian gravitation, there is a mathematical proof which is derived from accepted science and an example calculation showing that it makes correct predictions.

The ideas presented here have a significant, positive impact on Quantum Mechanics, with the removal of the gravitational force and the confirmation of a "flat space/curved time", background. Quantum Mechanics has been criticised for assuming a flat space-time background, another internal conflict within mainstream science. There are now ongoing attempts to marry the curved space-time of GR with Quantum Theory (Loop Quantum Gravity or LQG), but I believe we should be attempting the opposite, to reconsider the basis of GR from the perspective of a "curved time" only background, with time as the only operator.

General Relativity and Quantum Mechanics are brought closer together by the acceptance of these proposals and the quantum constructs developed today will, no doubt, remain intact. QM is not the main focus here but, later in the book, a proposed quantum theory of time explains many of the counter-intuitive aspects of relativity theory, in an intuitive, physical way. This new theory of time suggests a new explanation for the nature of subatomic particles and shows why the rules of gravitation become less applicable below the scale of classical laws and towards the Planck scale.

Inertia

No one knows what inertia is or how it works and there has been no explanation or a mechanism put forward, even in principle. Because we know enough about the rules of inertia to make predictions and to engineer with it, it is tempting to just accept it and to forget about how little we understand it. I therefore raise the fundamental question:

> *"Why should objects have **any** resistance to acceleration?"*

We can carry out calculations and invent rules about the equivalence of energy and mass and their behaviour, yet we still do not know what inertia is or why it exists. We know it is nothing to do with a body being in a gravitational field, and that the resistance to acceleration is directly proportional to its mass. Resistance is there, even in a vacuum in outer space, so we can at least deduce it is purely a property of the mass in question, a property of any and all mass.

This is *all* current science knows about inertia. In later chapters, an explanation for *why* energy is required to accelerate anything with mass is presented.

Recent developments have highlighted that it is the Higgs field that gives mass (or inertia) to matter and the Higgs field permeates all of space-time, it exists everywhere. It also has a non-zero-point energy. These are significant clues about the true nature of the Higgs field and its particle, the Higgs Boson.

The Bending of Light

Light is attracted by gravity just like any other entity, even though it is a pure energy beam with no mass. However, the behaviour of a light beam in a gravitational field has still not been satisfactorily explained since GR invokes the curvature of space itself to explain our observations. If Newton were alive today he might have said,

> *"That any test particle may have its path of motion curved through a vacuum by the physical curvature of nothing is to me so great an absurdity that, I believe, no man who has in philosophic matters a competent faculty of thinking could ever fall into it."*

This would have referred to the curvature of space but may also have implied that anyone with half a brain should never believe it! Today, too many scientists do believe this, without question, but I am suggesting we should at least be prepared to seek a causal explanation for this effect.

The transmission of light is via waves of energy which have no mass and therefore it cannot be accelerated in the mechanical sense. Whilst it may be true that an electromagnetic energy wave does "convert" into a photon[29] on impact with a solid body (at least for most wavelengths), this does not necessarily mean that we can consider the wave in transit to have a particle nature in the transit phase. Light can be viewed as either a wave or a particle, but not both at the same time. This wave/particle duality of light transmission is an unresolved conflict and we currently have no *satisfactory* explanation for the mechanism which bends light waves in a gravitational field. This issue is resolved later in the book.

General Relativity provides a space-time model which distorts space-time and consequently redirects light. The concept of the distortion of space itself is challenged here and we will take one step back from this misconception to explain how and why these gravitational effects occur and *how* space itself does *effectively* curve.

---

[29] Photon - The massless quantum particle of electromagnetic radiation

Energy Conservation

The total amount of energy in a closed system must remain constant.

This law of nature originates from the fact that we cannot create or destroy energy, but only change it from one form to another. For example, in the case of a rocket, chemical energy is converted to kinetic energy[30] and potential energy[31] plus some waste heat energy and other consequences of combustion. The sum of all the new energies when the rocket is in orbit still equals the sum of all the original energies before lift-off and the total energy within the system is what we term as "conserved".

In the case of gravity, we have invented the concept of "Gravitational Potential", which converts into kinetic energy as an object falls to the ground. In practice, this is a workable idea and much needed for our engineering. Nevertheless, we have created a convenient, limitless, arbitrary reserve of energy from which to draw our kinetic energy, and this seems impossible according to the laws of physics. If we choose to believe in Newton's force acting at a distance which breaks the principle of causality, then we have the problem of finding where the energy comes from, to exert this force in all directions, to infinity and indefinitely. We are thus driven to the creation of our "gravitational potential". Clearly, there must be another explanation, and relativity deals with this again with curved space-time providing the "cause and effect" for gravitational acceleration. This works in practice (and aligns with observation) *and* the mathematics of general relativity makes accurate predictions. Nevertheless, accepted science seems to ignore the simpler truth which has been under our noses since 1905 and even more clearly so since 1971.

Here we will establish the mechanism for gravitation which complies with all the requirements of current science, matches observation, conforms to accepted mathematical rules, resolves certain anomalies, aligns with at least one exact solution in relativity and yet requires no energy input for its application. It relies solely on the effects of relativity itself and the Causality Principle is strictly followed.

---

[30] Kinetic Energy (KE) – Energy of an entity due to its motion.
[31] Potential Energy (PE) – Energy of an entity due to its elevation in a gravitational field.

4 Space-time

To gain a complete understanding of gravity, we need to grasp the true nature of space-time. After all, space-time is the only medium available to affect the motion of particles travelling through it, directly. But, we must keep our feet firmly on the ground and resist the temptation to engage in the "flowery" imaginative thinking about space-time that seems prevalent today, in some scientific presentations in the popular media.

This "misrepresentation" of space-time may well be a genuine attempt to capture the general population's interest in the subject by "translating" some difficult concepts into everyday language and analogies. It may also be that some scientists themselves do not fully understand the nature of space-time and they, too, need the analogies to get some idea of the concepts. Whichever it is, I believe that the *accurate* presentation of such things to the general public is far more important than the concern that some people may not be capable of understanding them. We cannot afford to "dumb down" science in a futile attempt to educate everyone. This may sound elitist, or even politically incorrect but, if we are not careful in our attempts to popularise certain aspects of science, we risk confusing the very people who might, otherwise, contribute to real scientific progress, either from within (or perhaps even from outside) the scientific community. Ultimately, we may end up with only a small section of the population who really understand the scientific basics.

Certain beliefs, which form the basis of Special and General Relativity, are inexplicable, since no physical causality is given for inertial time-dilation and length-contraction in SR, nor for the gravitational time-dilation and curvature of space in GR. Although SR shows that lengths must contract and GR does show that space must curve in gravitational fields, no causality for these effects is suggested by the theories. I will demonstrate these causes later.

Johann Carl Friedrich Gauss (1777 – 1855), a German born mathematician, first presented the possibility for non-Euclidian, (curved) geometries. Such geometries do not have to adhere to straight lines and flat planes and they allow for geometric manipulations within a "curved" environment. This idea (and its associated mathematics) is a key feature in Einstein's deductive logic and mathematical constructs in General Relativity.

In one of his famous thought experiments, Einstein envisions the flat surface of a marble slab upon which are marked a set of straight, equally-spaced, parallel lines. More lines are then marked at right angles to the first set and with the same equal spacing to form a grid of squares. These lines are used as analogies for coordinates in two-dimensional space. To demonstrate the basis of GR, he then applies a heat source to a local area of the slab which expands the marble locally and results in the increased separation of the parallel lines in the region of the applied heat. The heat is an analogy for the effect of the presence of mass in three-dimensional (3D) space.

This two-dimensional slab surface can easily be envisioned in 3D by imagining similar, but vertical surfaces intersecting the original slab at right angles to, and being at the same positions and spacing as, the original coordinate lines. These 3D coordinate lines might then resemble a child's climbing frame. The heat is then applied at a local point within the marble "block" to produce the distortion of the frame within 3D space, the analogy for the "curvature" induced by the presence of mass.

As soon as you allow for the possibility of the curvature of 3D space, this opens up a whole new field of geometry. The geometry will inevitably work and produce accurate predictions, since the known rules of physics are inbuilt within the manipulations. It is in the nature of mathematics to always remain sensible, so long as you stick to the rules. Nevertheless, this does not necessarily mean that space really can be curved just because we have allowed it to be in our mathematical treatment of it. The mathematics of GR is an abstract, geometric interpretation of events within space-time and is highly successful in making accurate predictions. Even so, contrary to some claims, GR does *not* identify the physical cause and effect of gravitation. It does not say *what* the "heat source" is within the marble slab.

There has never been a proposal for *how* the presence of mass distorts 3D space. Although we can understand the analogy of heat expanding and distorting a physical material, there is no known cause and effect for the distortion of a vacuum, the curving of nothing, yet the very foundation of GR relies upon the unquestioning acceptance of this proposition. The basic premise of GR is therefore a self-fulfilling prophecy, a somewhat "circular" argument, based on the belief that the presence of mass *somehow* curves space as well as time. If we choose to believe that mass curves space, then we inevitably end up with GR. If we choose *not* to believe it (and I suggest this is a far more realistic choice), then we must motivate ourselves to find another explanation for the observed effects close to massive objects in space, despite the fact that General Relativity accurately predicts these effects. I know this is hard for some professional physicists, but minds should still be open on this issue, as I am convinced that no-one can be entirely comfortable with this abstract curvature – the bending of nothing without any cause. I know the mathematics of GR does show that this curvature is inevitable, but the math does not get to the root cause. Later in the book, I demonstrate the physical causality for the relativistic effects of inertial time dilation, inertial length contraction as well as the effective curvature of space in strong gravitational fields. The phenomenon of time is identified as fundamentally different from the three physical dimensions and there is an objection to the equal treatment given to all four "dimensions" within the philosophy and mathematics of GR. There is practical proof of all three of these effects, but I will show that the variation in time rate is fundamental and that it is the cause for the other two effects. In fact, I will demonstrate that time is the fundamental cause for *all* effects.

Velocity is *distance moved* per *unit time*. It is the fraction, distance/time and so if the unit of time becomes different then so does the velocity. A change in velocity is defined as acceleration and so we can quickly deduce that acceleration must be imparted to any object in a field where the time rate varies (curvature of time) and without the application of any force. If you think clearly about this, you will see that such an acceleration induced in this way will be independent of the mass of any object being accelerated, since it is a purely temporal cause which does not involve any force (the effects are independent of mass). Inertia simply does not come into it. This is the fundamental "cause and effect" of Newtonian gravitation and this conclusion agrees with the mathematics of GR (for weak fields). Note that we have just arrived at the same result, in principle, from a different perspective, using only one paragraph of deductive logic.

GR concentrates on observational effects and, although it recognises Proper lengths and Proper time, it devalues these realities within any frame of reference. The proposals within this book give equal credence to reality and relativity and they focus on the realities, as much as the relativistic issues. But, when there is a conflict between reality and relativity, reality must always prevail!

General Relativity views the three dimensions of space as non-Euclidian[32]. Euclidean space is flat in all three dimensions and cannot be distorted. Einstein's non-Euclidean space-time means that space itself, as well as the time "dimension", can be curved (vary according to position). A practical justification for this idea is the observed bending of the path of a light beam in a gravitational field. This observed deflection of light gives support to the idea that space itself must be curved by the presence of mass, to deflect light in this way, but we will see later that this curved path or geodesic is an effect of the way time changes with position and that this time curvature not only causes Newtonian gravitation directly, but is also the root *cause* for the effective curvature of space.

I assert that space or volume must be, by its very nature, Euclidian and "flat" in the three physical dimensions. I propose that it is the time rate only that can vary within its field, (can be curved within space-time) and that it is this time rate variation alone which must cause the bending of the paths of motion of any entity, including light, in both strong and weak gravitational fields. We shall see later how this can be and that the effective curvature of space is merely a secondary effect from the extreme curvature of time across distance, in strong fields.

The Euclidian model of space-time is known as Minkowski space-time, or "Flat" space-time, and was the space-time model proposed by Herman Minkowski in conjunction with Special Relativity in the early twentieth century. Herman Minkowski (1864-1909), was a highly regarded scientist, a German Doctor of Mathematics, born of Jewish, Lithuanian parents. He taught at the

---

[32] Euclidian (adjective) – Euclid was the Greek mathematician who invented geometry and his name is thus applied to the geometry of flat planes.

universities of Bonn, Gottingen, Konigsberg, and Zurich and one of his students was a certain Albert Einstein.

The idea of a flat space-time is partly accepted today by many relativists and is typically presented as: "Newtonian gravity as the curvature of time", as in the book, "*Gravity from the Ground Up*"[33] by Bernard Schutz, to give one example. However, Schutz, and others in the scientific community, seem to present this idea as merely one way of looking at gravitation and that gravitational *red shift* is a consequence of the gravitational field. I am saying exactly the same thing in terms of how gravity relates to time dilation[34] but, instead of asserting that gravity causes time dilation, I propose that it is the other way around! I shall prove that time dilation is not a result of gravity but that it actually *causes* our gravitational field to exist. In other words, gravity *is* time! More accurately, the gravitational field is the time rate field!

The non-Euclidian (curved) model of space-time and its associated mathematics is accepted today, perhaps partly due to the concern that if the physical dimensions of space were to be held constant, the variation or curvature of the "time rate" on its own, might not come up with the right numbers to match observation. This concern is born of a lack of understanding of the effects of the time dilation field (time rate field). Further on in the book the real cause of the deviations from Newtonian predictions in strong fields is demonstrated (and a different way to understand GR is proposed).

Maybe it is also the case that we have become confused by the idea of space-time and are unable to clearly visualise it, and therefore to represent it as it really is.

Perhaps we are too much in awe of the established and complex mathematics of General Relativity and are thus dissuaded from questioning the idea of the curvature of space-time and of space or volume. We have failed to challenge this idea and have thus far avoided having to confront the impossibility of the curvature of "nothing".

I take the "realist's" view, in the sense that although I understand and accept the idea of a space-time continuum, I nevertheless believe we should consider space or volume as an

---

[33] "Gravity from the ground up" by Bernard Schutz – ISBN 0 521 45506 5, first published 2003.

[34] Time dilation – The slowing down of the rate of passage of time.

entirely separate entity, from time. Here therefore, is my understanding of the following:

Space, The Vacuum and The Void

Quantum Mechanical Theory recognises that there is more to the vacuum than simply empty void. It has "vacuum energy", it can be polarised, it is unstable and it is a "sea" (Dirac Sea[35]) of potential, "negative" antiparticles. There are different fields in space-time from which different particles emerge under extreme conditions. Even in completely empty space, virtual particle and antiparticle pairs continuously come in and out of existence - if only very briefly. Nevertheless, all these attributes apply to the vacuum nearer the quantum scale and not the cosmic or macro scale. The vacuum at the macro scale, above 10^{-20} metres, from where the classical rules of physics apply is, essentially, nothing. It is, to all intents and purposes, an empty void and since we are investigating the nature of gravitation, which operates at the macro scale, then we will also consider the vacuum at the macro scale as an empty void.

To make it easier to accept the notion of complete emptiness, let me clearly differentiate between the "vacuum" and the "void". If we are going to label "empty" space as the "vacuum" with all its attributes and properties, then we must still envision the "void" within which this vacuum exists. You simply cannot get away from this principle that, whatever we observe to be in the vacuum, there still must be the empty void within which the vacuum resides. Of course, this process of reduction ceases when you get down to the void since there is nothing left to reduce. Quantum Field Theory suggests the vacuum consists of fields which affect the behaviour of particles, as well as providing the energy to produce the particles themselves. Later in the book, you will see that there is likely to be only the one field which results in all these effects and that the many particles are simply the result from different energies of impact on the one field.

---

[35] Dirac Sea - Paul Dirac (1902 - 1984), renowned quantum physicist discovered antimatter and that there must be a sea of potential "negative" particles from which antiparticles emerge.

Whatever entities are present in the vacuum, we must admit to the possibility that they do not *have* to be there. In such hypothetical circumstances, when nothing is present, we are inevitably faced with a true void. This situation may or may not have been the case before the Big Bang or whatever the beginning was, but we have to accept the notion of the void as just one part of "the vacuum" in addition to all its other attributes. We will see later how all these attributes of the vacuum can be reconciled with an absolute void but for the purposes of this part of the book we can decide the following:

A cubic metre of space is a cubic metre of nothing, despite vacuum energy, quantum foam and other mathematically derived phenomena and if you keep removing entities, real or imaginary, from physical space you will inevitably arrive at absolutely nothing. It is impossible to deny the "existence" of the void, the possibility of absolutely nothing, since this is what was there before anything was put into it. Time, on the other hand, occurs everywhere in the vacuum, at all scales, and the ultimate conclusion is that, fundamentally, there is only the nothing of the void and the passage of time throughout it. Fundamentally, there is only space-time.

Our imaginary cubic metre will always measure 1 metre x 1 metre x 1 metre wherever it is, however fast it is "travelling" in a moving frame of reference, whether it is in a gravitational field of any magnitude (and however distorted it may appear from the frame of reference from which we might be considering it). Even in very strong gravitational fields, where one corner of the cube looks distorted when viewed from the opposite corner, it is still one cubic metre when summing all the proper volumes within the cube. The "Proper" lengths of its three dimensions, the lengths as measured within the frame of the cubic metre (or the sum of all its proper lengths) never change whatever the circumstances. Even Relativity theory states this to be the case. "Proper lengths" in both SR and GR are recognised as constant *within* any frame of reference and that is the definition of the term, "proper".

Our cube is completely empty... well, almost. In practice, there are usually a few atoms, maybe some electromagnetic radiation passing through it, some residual gravitational field (yet to be defined) and of course the three-degree background radiation[36]. Surprisingly, when you have removed absolutely everything from the vacuum, it still "weighs" something. I take this to mean it has some energy of its own and we will see later that this "zero-point energy" is nothing more than the effects from the passing of time.

However, just because we can never get down to absolutely nothing anywhere within the observable universe, this does not exclude the idea that we first require the absolute nothing of the void before we can put something into it. So, for the purposes of our clear understanding of space, we will consider a hypothetical void, a total vacuum shielded from all radiation sources and remote from any gravitational field. Even the three-degree background radiation does not mean we cannot have a totally empty "void" to start off with, since the void may have been potentially available (I hesitate to say "existed") before the Big Bang.

Our cubic metre has no attributes. Even its boundaries are imaginary. It is incapable of having any attributes because it is by definition, *nothing*. Any attributes of the vacuum are due to something else within the void, not from the emptiness of the void, but from the properties of the fields and particles within the vacuum. In particular, the void has no directional properties and cannot be biased in any direction. It is the same at any position and in any direction. It cannot be changed since there is nothing *to* change. It cannot be curved or otherwise distorted since there is nothing *to* bend, curve, expand or distort. Protagonists of the directionality of the void and of General Relativity must necessarily argue against this view, but it is the other entities within the void that might give space any directionality and not the void itself. The basis of General Relativity requires the curvature of space to make the theory work. Conversely, it is initially envisaged that space can be curved and the mathematics is generated from this fundamental, yet unreal assumption. The idea that space itself can be curved was an assumption without causality, a guess, despite the success of the mathematics.

Any attempt to give the void some substance or to create a new entity, such as the fictitious "Aether", or some fabric, often referred to as "The fabric of space-time", or even the "quantum

---

[36] Three-degree background radiation – Residual heat from the Big Bang.

foam" (another unproven entity), or any other attribute that we might invent, is self-deception. These creative, yet false, ideas will only serve to distract us in our attempts to understand the physical world. The "Quantum foam" is a hypothetical notion emanating from the combined mathematics of Quantum Mechanics and General Relativity yet there is no observable evidence for its existence. The "foam" remains just a postulate, so far unproven by experiment.

Conventional scientific philosophy maintains that space must be "something" for it to contain anything and for it to have any effect on events, but this philosophical belief lacks any scientific basis. In any case, I am advocating that the void can never have any effect on events since it is, you've guessed it, *nothing!*

So, our first philosophical disagreement with General Relativity is our assertion that space or volume cannot be distorted but that space-time can be, effectively anyway.

A flat space-time is also assumed by Quantum Mechanics as the basis for that branch of science and Cosmology also accepts that space is essentially "flat" at the cosmic scale. There is therefore a long-standing conflict within the scientific community on this issue quite apart from these new proposals.

The well-known two-dimensional gravitational analogy in Figure 1 shows a grid, distorted due to a heavy object resting on it like a medicine ball on a soft mattress or trampoline. This is a useful tool for visualising the rules of gravitation, but it should be realised that the bending of the grid lines is analogous to the effects of the gravitational field induced by the massive body and not to the distortion of space itself. Space, or nothing, cannot be distorted. The downward slope of the gridlines represents the direction of the attraction being always towards the centre of the body, and the degree of slope, or gradient of the gridlines represents the *intensity* or strength of the gravitational field at any point. The steeper the slope, the stronger the attraction and the closer we get to the body, the stronger its attraction. In this sense only, is this picture an analogy of a gravitational field.

More than anything else, this particular analogy serves to show how analogies in general are less than ideal for demonstrating the realities of our world. They never fully describe reality and are merely aids to help us understand *some* of the principles involved.

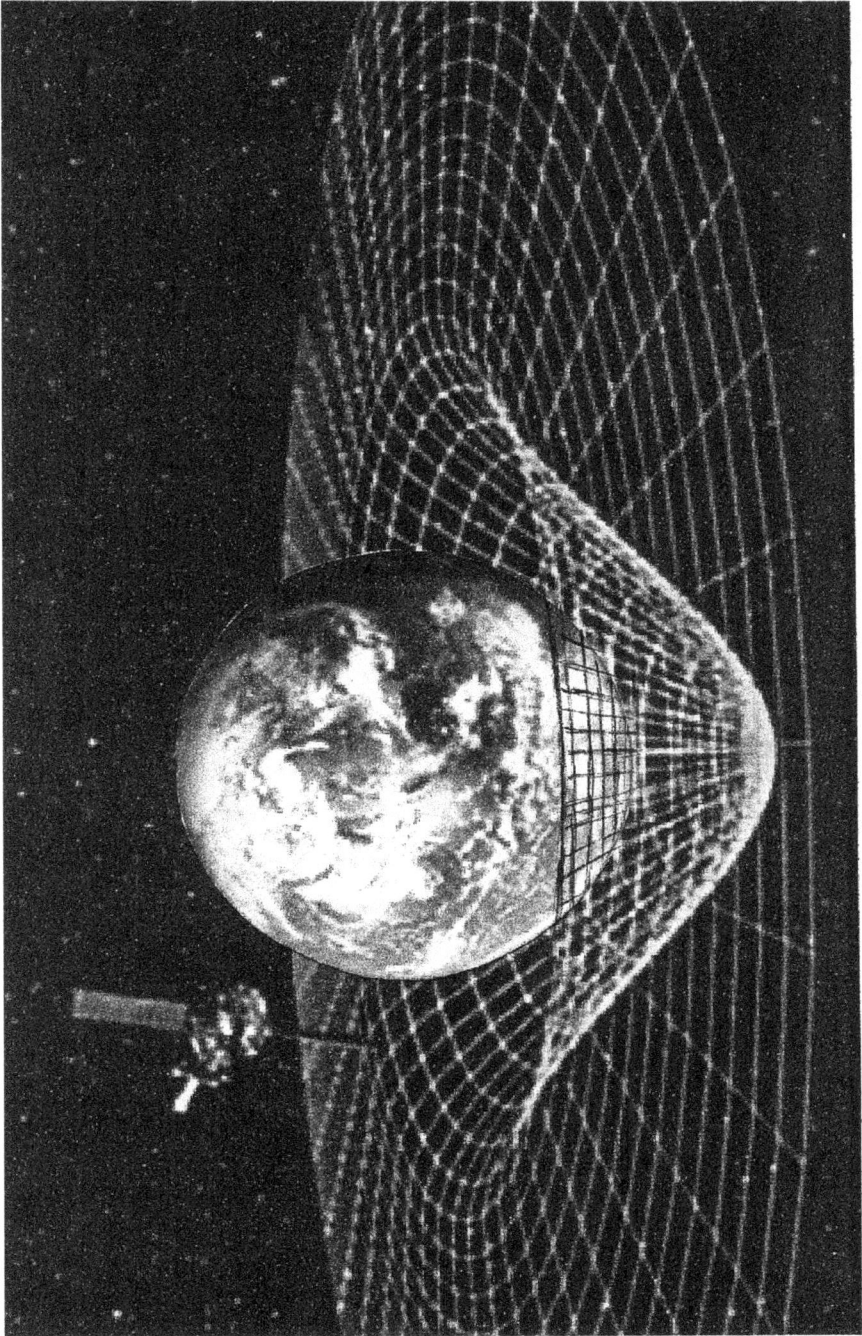

Figure I: Gravitational analogy

Space-Time

Space, the void, (or volume), and anything within the void, cannot "exist" without the passage of time. If time does not pass within the void, then space never did exist, it does not exist now and it never will exist because there is no past, present or future. From this understanding we can see that the passage of time actually creates the void and the way in which time creates space will be demonstrated later.

This is why we regard time as being the fourth dimension. It is the fourth requirement for "existence". Actually, I would prefer to call it the second dimension since the three dimensions of length, taken together, can be regarded as just one entity and there are only the two fundamentals – the void and the passage of time. It is more commonplace though, as well as being mathematically practical (in Cartesian coordinates), to use the three physical dimensions and the fourth dimension of time so I will not press this notion again. In fact, if you take the void as the inevitable, infinite nothingness then you must conclude that only the entity of time was necessary to create our universe with everything in it and we will explore this view later in the book.

Just because we label time as a "dimension" does not make it the same *type* of entity as the other three physical dimensions. Time is a real phenomenon, or series of events, whereas the three physical dimensions are not, they are merely potential, the potential to contain something and this potential cannot be realised without the passing of time.

So, the pre-requisites for existence are the three physical dimensions of the void, and the passage of time. In this sense, space is undeniably linked to time and we can call the combination "space-time". The proposed theory shows that it is only the rate of passage of time which can vary with position or with speed and that this change of time rate with position is the root cause for the effective curvature of space.

If we accept this notion and please *do* try and accept it, if only temporarily, then this presents us with a universe of a variable time rate field in a Euclidean "flat" space. I am advocating this view of space-time based only on "proper" lengths or dimensions and not on relative observations of lengths or dimensions. I am giving *realities within* frames precedence over *observations between* frames, regardless of whether these realities and observations are derived from thought experiment or real experiment.

Time rate variations are caused by either the greater speed of an entity relative to a "less accelerated"[37] frame of reference, or by the proximity to a massive object with its varying field of time dilation. These examples of time rate change are taken from Special and General Relativity and are a key part of accepted science. The new theory also relies upon them but in a subtly different way as you will see.

Within Space-Time, space and time are inextricably linked as requirements for existence but they are otherwise entirely separate entities and should be treated as such. In fact, I have proposed that space is not actually an entity at all, because it is simply nothingness. It is the nothingness that was there before the beginning of time and it will still be there after the end of the universe. It requires no creation for its "existence" because it is nothing, and this leaves us with time as the *only* entity, the only *reality,* that exists within the void. The fact that a few atoms and a small amount of radiation may also be present does not undermine this argument. Even effects attributable to the quantum "foam" are likely to be the effects of the nature of time itself near the Planck[38] scale and so it cannot yet be argued that there is no perfect vacuum or that there is no possibility of absolutely nothing. At best it might be argued that this is impractical and there really *is* no perfect vacuum anywhere, but I maintain that the notion of the void is much less abstract than the curvature of it.

This idea does not conflict with Einstein's theory of Special Relativity which states that time, length (and therefore volume), as well as mass and energy, are all relative depending on your frame of reference.

---

[37] "Less accelerated" frame – Either, the frame of reference initially shared with an entity which has accelerated and increased its speed, or, a frame of reference higher up in a gravitational field than the more accelerated frame. For convenience, the terms "Moving" and "Stationary" are sometimes used in future text to mean "More accelerated" and "Less accelerated" respectively.

[38] Planck Scale – Max Planck, 1858-1947, a German theoretical physicist who originated Quantum Theory. The Planck scale is down at the smallest fundamental units of length and time.

Special Relativity does not state that your *actual* (or "proper") length compresses to zero at the speed of light for instance, but more that from a "stationary" observer's point of view it *looks* as if you are compressed flat. This is simply due to the geometry of relative motion and the fact that light propagates at a maximum, finite speed, "*c*". This physical compression is therefore unreal to the person travelling at the speed of light. Admittedly, this is the extreme case but, even at moderate speeds of a small fraction of the speed of light, mass effectively increases and length effectively shortens in the direction of motion, both imperceptibly at low speeds, but these effects are purely relative between frames and are not real within either frame.

We will be considering the effects of travelling at the speed of light, even though it is impossible for any entity with mass to attain this speed within the lifespan and energy limits of the universe. This licence is necessary to allow us to see more clearly the trends of the effects of speed by considering the hypothetical effects in the absolute limit, but if you prefer, you can simply imagine yourself as a photon.

Reality is only a valid concept *within* any particular frame of reference and different frames of reference each have their own unique realities.

The reality *within* a moving frame of reference is that space, mass, energy, volume and length remain as they have always been and everything remains normal, even as you hurtle through the universe, approaching the limiting velocity of light. All relativity means is that the length of something in a frame of reference moving at the speed of light *seems* to compress to zero, when viewed from a stationary perspective.

Likewise, lengths in the stationary frame also seem to compress when viewed by the moving observer, but we know this cannot be real.

Space only *seems* to compress.

Mass only *seems* to approach infinite mass, but......*time*......is different!

Time is the exception since clocks really do work at different rates between different frames' in relative motion.

Time dilation is *real*.

This is demonstrated when a clock which has been in the moving frame comes to a halt. Although the traveller's dimensions and mass are the same as they were before the journey started and have been so throughout his journey, despite the relative observations to the contrary, his clock is now slow compared to a clock which had remained in the stationary frame. So, the travelling clock really *has* been operating more slowly for the period of the journey due, simply, to its relative speed. So, the view of this clock from a stationary frame during its travels is not just relative, but is in fact also, *real.*

The physical changes, however, are purely relative observations and are not real. They are illusions or distortions of reality. The correct ontological, philosophical terminology is that length contraction is ontologically subjective, but time dilation is ontologically objective. Time dilation is a real phenomenon, which is evidenced by the time lost on the moving clock, but length contraction is simply an observational effect, analogous to the distortion of an object when viewed through an optical lens. The image shows a real distortion, but we know the object remains the same shape behind the lens.

Before we proceed further, we need to note my use of the word "real" and in this context I propose the following definition:

An event, occurring within its own frame, which causes a permanent effect when entering a different frame, must be "real". An event which can be observed from a different frame, but which has no effect when entering any other, different frame, is not real and is purely relative.

In conclusion:

Only real events can have an effect in a different frame

and of course,

For any real effect, there must be a real cause

This fundamental difference between the physical dimensions and the "dimension" of time itself is not recognised by the theory of relativity and this way of looking at time is a challenge to current science.

I am certain this will have been raised previously, but mainstream science must have always rejected it. Once accepted then taken further, it has a significant impact on our understanding of relativity and of the nature of our world. However, for those in fear of the impact on science, I can assure you, there is no need to panic.

The classic example of motion induced time dilation is the hypothetical round trip to our nearest star, Alpha Centauri, which is about four light years[39] away. If you were to travel there and back at the speed of light then, for the sake of simplicity (and assuming you could instantly accelerate and decelerate to and from this speed), it will have taken eight years for you to get there and back as measured by an Earth clock, but your moving clock has stood still for all that "Earth time". In other words, you got there and back in an instant, only to find that eight years have passed on Earth.

Clearly you cannot have instantaneous acceleration or deceleration and in practice there will be a significant departure from this effect due to the time taken to get up to (and slow down from) the speed of light, even if it were attainable. Nevertheless, the simplest, if impractical thought experiment best demonstrates the point.

This is the extreme case of time dilation since time slows to a halt at the speed of light but whatever speed we travel at, our travelling clock is in fact running slow relative to a clock which has not moved, albeit imperceptibly at speeds much slower than the speed of light.

This is the only reality from relativity.

The length distortion and the mass increase (to infinite mass at the speed of light) are merely illusions and our interpretation of the mathematics, being governed by geometric effects but exclusive of temporal effects, does not reflect reality.

---

[39] Light year – the distance light travels in one year at a velocity of 186,000 miles per second, or 300 million metres per second.

The following is a fundamental principle upon which the New Theory is based:

> *Only time can vary with position or with speed and this time dilation is absolute, it is not just relative. The physical distortion observed between moving frames of reference or within a gravitational field is purely relative but the time dilation is real..*

But what about mass increase? Clearly your "proper mass" can never change and within the accelerated frame your mass is the same as it was before you started to accelerate, even as you approach the speed of light. So, why does the math tell us your mass is approaching infinite mass? Well, in the same way that the math can tell us your length has contracted when this effect is really due to time dilation, then so it is with mass increase.

As you accelerate towards light speed, your clock slows noticeably and approaches zero time rate as you approach "c". This is the only real occurrence within the moving frame and it results in the relative effects of length contraction and mass increase. These are purely relative observations between 'frames. The math tells us only that it is becoming impossible to accelerate further but it is the time dilation which is the cause, not any real increase in mass. Mathematically though, it is "as if" mass has increased and so the mathematics will offer us this option to explain why it is becoming more difficult to accelerate. Mathematics will always give us the right relationships but we should never rely on it to give us the right philosophy. *We* are the Masters of Philosophy, not mathematics.

If you are approaching the speed of light and still have your rocket engine blasting away at full throttle, then you are chucking mass out the back at a very high rate, say 10,000 kg per second (never mind there is insufficient energy in the universe for the moment, let us look at the principle). The reaction to this, in accordance with Newton's second law of motion, F = ma, is what propels you forward and you accelerate. You still feel the same kick in the back and you are indeed accelerating as before no matter how close to the speed of light you get. This is demanded by the principle of relativity which requires that your speed does not affect the laws of nature within your frame of reference.

However, if I am standing still, observing the rate of mass discharge from your rocket engine, then since my clock is running as normal, but your clock is almost at a standstill, I will see a pathetic rate of mass discharge of say one gram per year (my year) and this is the reason why it will become impossible for you to accelerate further, *but only as observed from my frame of reference.* It is *as if* your mass has increased relative to me but it is the time dilation which causes this effect and we misinterpret what the math tells us. Our interpretation of the math is "confused" by the different frames of reference. We see any acceleration force as being applied from the static frame resulting in negligible acceleration and we draw the conclusion that mass has increased. But, in reality, the force is applied from within the moving frame and because of the slower clock there, acceleration is normal and in accordance with Newton. The math thinks you are throwing rocks at the rocket in order to accelerate it but, in reality, the rocks are being thrown out the back *from* the rocket.

So, what *is* real? Well, reality is only ever relevant *within* a frame of reference and there are two realities to consider here. One is the reality in the moving frame and the other is the reality in the stationary frame. Cross-frame observations are never completely real. From my stationary frame it will take you an inordinate length of time to accelerate further only slightly, but this is due to my relatively faster clock and your relatively slower clock. From your moving frame you will indeed continue to accelerate as normal, since the laws of physics are unchanged in the moving frame but only in relation to your clock which by now is almost stopped, relative to me. You are "experiencing" time normally but your time rate compared to the stationary clock is now very slow, indeed.

Your speed and acceleration, which are both time related events, are *red shifted* when observed from the less accelerated frame, but, in your moving frame, you will observe a temporal *blue shift*[40] of the rest of the universe passing by through your

---

[40] Accepted science takes the opposite view and *red shift* is predicted when observing the less accelerated frame. This is being challenged.

porthole.

As far as you are concerned, you are still accelerating and, in theory, will reach the speed of light, since every little increase in speed (produced by the reactive engine forces) produces increasingly more time dilation, which in turn amplifies this speed increase, relative to the static universe. This amplification effect increases ever more greatly the closer you get to light speed and you will get ever more easily drawn faster and faster until you finally "pop" into the speed of light and zero time. From my stationary perspective though, you will never make it. Indeed, by the time you do make it, the rest of the universe will have come to an end! It is not that there is insufficient energy in the universe to propel you to the speed of light, it is more that there is insufficient *time*. Later we will see that time *is* energy, ultimately it is the *only* form of energy. Of course, this is purely theoretical since you have some mass and so can never reach the speed of light. You have to be without mass for light speed to be a practical proposition.

The practical problem with near light speed travel is that when you have reached "c", you have totally lost control over where and when you stop. Photons cannot decelerate, they just head toward the edge of the universe until they hit something. So, the challenge will be to calculate and control the speed and time at which you must decelerate – if you wish to end up somewhere near the planned destination. Clearly, this maximum speed must be less than "*c*" to have any control whatsoever and to avoid the end of the universe, but as close to "*c*" as possible whilst maintaining enough control for sensible travel. The calculations involved must consider the "curve" of acceleration, the duration (at whatever your maximum speed is) and the deceleration "curve". The integral of these will determine your total distance travelled and the proper time to achieve it. It will be the onboard clock that will rule your deceleration event. The closer to light speed you have been travelling, the more accurate the clock time and breaking point will need to be. Even if light speed could be achieved it would be of dubious benefit to the traveller since he leaves everything behind him temporally and consigns the present universe to his "history". He should only embark on such a journey with the knowledge that it is effectively a one-way trip. Nevertheless, the point is worth making, if only to get the facts straight.

We must not believe that our mass increases or our lengths contract due to our motion and we must understand that these false ideas are simply due to the temporal effects of inertial time dilation. They are distortions of reality, from observing across two frames of reference, like looking through a lens and seeing optical distortion. It is only the time dilation and its effects on time related events that is real.

Time

We will now consider, in more detail, why the time "dimension" is different in nature to the three physical dimensions, but first a clarification of some basic terminology. When we talk about "reference frames", or "frames" for short, we refer to the reality within our stationary environment as being the "stationary frame", and we refer to the reality within our moving environment (say on a jet or a space craft), as being the "moving frame". By the term "reality", we mean the physical environment within the frame, including, length, mass and all the laws of nature.

The following points highlight the fundamental differences between time and space. Some boil down to the same thing, but it is as well to highlight all aspects from all points of view.

- Firstly, the physical dimensions are not entities, but simply dimensions (void), which are available for any entity to exist within. Time however, *is* an entity or phenomenon which occurs throughout the physical void and passes at one rate or another depending on local circumstances.

- The limits of the three physical dimensions are between minus infinity and plus infinity in any direction. Time has its limits between zero in the extreme accelerated reference frame, and the time rate displayed on the remote, stationary, theoretical fast ticking clock at infinite distance from any gravitational field. The latter is the fastest time rate possible, the "universal" time rate, or "universal time". Motion or the presence of mass can only serve to slow or dilate this.

- Dimensions can be infinite whereas time is finite, (between zero and universal time). Time is said to pass at the rate of 1.855×10^{43} Planck times per second and so, clearly, the time rate is finite. At light speed or at the event horizon of a black hole time has literally stopped in absolute terms and all events, relative to the rest of the universe, cease.

- Time is unidirectional or asymmetrical in that, although there is a past, you only move through it in one direction, toward the future. Physical dimensions can be travelled into, from any direction towards minus infinity or plus infinity. A physical dimension is therefore bi-directional or symmetrical.

- The three physical dimensions are passive in that they are merely available for us to exist within. Time though, is active and drives us into the future as if we are on the crest of the wave of time. Space cannot affect anything, but time is a phenomenon with affects everything.

- We are driven toward the future at a rate dictated by the time rate in our frame of reference but we can move in any physical dimension at whatever rate we choose depending only on the energy we apply (although a change in velocity does change the time rate and therefore the frame of reference).

- The time rate is fixed in any frame, depending on the relative speed and/or the elevation in a gravitational field. However, this does not mean that a change in time rate will distort space, but only that the frame of reference will change. Dimensions are unaffected within all frames (proper lengths never change).

- The existence of any entity does not require movement through the physical dimensions, but it does require movement through the time "dimension". Conversely, it requires time to "flow" or pass, for the entity to exist.

- Movement through space is fundamentally different from movement through time. We can actually move through space, but we do not move through time. Rather we are swept along by it, with the rest of the universe, although there are slight variations in its rate of progress, depending on our location and our speed.

- We can maintain our position within the void and remain "static", we can choose to move forward in any direction or backwards in the opposite direction. We have maximum degrees of freedom. We can never do this within the time "dimension". We exist only in the present, not in the past, nor in the future, but always in the flowing, fleeting moments of the progression of time itself. We have zero degrees of freedom.

- When we look into any physical dimension, in any direction, we can "see", theoretically, to infinity, limited only by the speed of light and the resulting time delay for the light to reach us. When we look into the time dimension, we cannot see at all in either direction. We are blind. It is only with our memories, or records of past events, that we can try to recall or see our history, but recollection is not observation. Remembering is not seeing. Nor do we see into the future and we are all temporal "flatlanders" in the present.

- With time there is uncertainty. There are numerous possibilities and even probabilities for the future, but nothing is absolutely certain. The past *is* certain since the present would not be what it is, if the past had not occurred. With space, there are no possibilities, only potential, all-encompassing, yet blank, without any certainty, probability or possibility... until time starts to pass.

- Length, or distance, is an abstract, relative notion, whereas time is not. We measure one metre as the distance travelled by light in 1/300 millionth of a second. One light year is defined as the distance travelled by a beam of light in one year. We always have to make comparisons. Even the Planck length[41] is defined simply by the distance light

---

[41] Planck Units – First identified by Max Von Planck, (1858-1947). They are the fundamental, natural units of space and time and are labelled – Planck length and Planck time. There is also a Planck mass together with all other physical quantities.

travels in one Planck time. But, the Planck time is definitive. It requires no comparisons and has a finite, absolute value to which everything else must be compared. We will pursue this later and discover the true nature of time itself.

- For existence, continual change is required. Whether it is the rotation of subatomic particles within atoms, the orbiting of the Earth around the Sun, or any event with the passing of time itself, there is always change. The entity which brings about change is time. Without time there is no change, no evolution, no existence.

- Time slows down with increasing velocity, due to inertial time dilation. Its rate of progress is inhibited by the motion, in accordance with the Lorentz transformation. At the speed of light, time stops altogether and distances become meaningless. Time also slows in the vicinity of large masses, due to gravitational time dilation. Since velocity is a form of energy, kinetic energy and mass is also a form of energy, mass energy, then these effects on temporal progress suggest that time itself is also a form of energy, energy that can interchange with kinetic energy, or be reduced by the energy contained within a mass.

Because of all these fundamental differences between time and the physical dimensions, we cannot simply assume that they are the same *type* of entity and that they will all be affected in the same way by events in our universe.

> *Time is not a dimension. It is an objective phenomenon throughout the void – a physical process*

These are the reasons why the new theory is not just content but is *compelled* to treat time differently from the three physical dimensions. This idea is in opposition to General Relativity which allows the general case, the case of any or all four dimensions being potentially curved and of all of them being linked in a four-dimensional Gaussian coordinate system. GR treats time as just another dimension!

We should note at this point that geometry is indeed an appropriate tool for predicting certain effects on reality, for example, to prove that the time rate varies relatively with relative speed. Purely geometric solutions though, are unreliable, since the dimensions in geometry are bi-directional but time is unidirectional and any logic employing both can easily become confused. Length contraction is a case in point.

At any speed, time is said to slow down and lengths are said to contract in the direction of travel and, at the speed of light, time stops and lengths contract to zero-length.

These effects are borne out by experiment, but why do we need two causes, when either one of them produces the predicted effects on its own?

If lengths really have contracted to zero-length, then we do not need a stopped clock to get anywhere we like in zero time. If time really has stopped then there is no need for lengths to have contracted to zero-length to give us the same effect. Clearly, only one of these is real and the other just a resulting, relative illusion.

We need to use our powers of reason and visualisation in conjunction with geometric solutions, carefully, to enable us to identify what is real and what is purely relative.

Time Dilation

Motion induced time dilation (Inertial Time Dilation)

Special Relativity shows us that the passage of time is slowed down for a body or particle, or beam of light (or any entity for that matter), when it moves, relative to another "stationary" (less accelerated) frame of reference. There are infinite frames of reference, none of them stationary in absolute terms (allegedly), but two is the simplest number to consider when looking at basic principles.

Special and General Relativity are continually being validated and we can see the proof of motion induced and gravitationally induced time dilation, today, from our activities in space exploration and from the corrections which are necessary to keep the global positioning system (GPS[42]) in operation.

---

[42] GPS, Global Positioning System – A space-based global navigation satellite system giving reliable location and time information anywhere on or near the Earth's surface

The GPS needs to take account of both gravitational time dilation and inertial time dilation to avoid a disastrous build-up of error. Although the differences in time rates between ground and satellite clocks are extremely small, they nevertheless accumulate over fairly short time scales to such an extent that the system quickly becomes inoperable.

The demands on the accuracy and synchronicity of the clocks in the system are very high.

An object rotating around, or otherwise moving relative to, another "stationary" (less accelerated) object has a slower time rate than the stationary object because of its relative motion. Consequently, the clocks on board the GPS satellites need adjustment, firstly because they *lose* time due to their motion relative to a clock on Earth and secondly, because they *gain* time due to the increased time rate at orbital elevation in the Earth's gravitational field. They need to be adjusted and pre-set before launch, firstly with a slightly faster time rate than the "stationary" Earth clock, to compensate for the slowed time rate when at orbital velocity. This motion-induced time dilation is in fact overpowered fivefold by the *increase* in time rate of the satellite clocks, due to their higher elevation in the Earth's gravitational/time dilation field. The net effect is an *increased* rate of the passage of time for the satellites when, in orbit, because the clocks run faster for the satellites compared to the identical Earth-bound clocks. So, the net pre-launch adjustment is in fact a slower time rate setting such that, when time speeds up when in orbit, the satellite clocks then run at the same rate as the earth-bound clocks. Nevertheless, the *slowing* of time induced by the *motion* of the satellites is indeed real and has to be compensated for in the overall, net, pre-launch correction. This net, pre-set adjustment synchronises all the satellite clocks with the ground-based clocks and ultimately avoids errors in distance with, say, a car's GPS navigation.

With the extreme accuracy of atomic clocks, motion induced time dilation can even be measured when comparing a clock on an aircraft with an identical, stationary, synchronised clock on the ground and this experiment was carried out by Hafele and Keating in 1971 as a test of Special Relativity. An atomic clock was flown around the world in both directions. The travelling clock showed a slight reduction in overall lapsed time compared to the previously synchronised, stationary clock when the journey was over. Although this difference in recorded time was very small, nevertheless, it was real and measurable, and the results confirmed Einstein's predictions from Special Relativity.

From the Hafele & Keating experiment we must conclude:

> *Time dilation due to motion is a proven, real effect, for the entity which accelerates and moves, relative to its initial, less accelerated frame.*

This clear result from the Hafele & Keating experiment seems not to be fully understood and its ultimate consequences have been avoided by mainstream scientists. The mainstream asserts that time dilation is a purely relative effect and is no more real than length contraction but when time stops at the speed of light, this must be absolute. It stops relative to *everything*. Because of the overriding belief that all motion is purely relative and that there is no preferred reference frame, they cannot accept the reality of time dilation since this would ultimately mean there *is* a preferred reference frame. So, they have to view this effect as being somehow unreal and time dilation has been given the same status as length contraction, an illusion. But, the final time difference from the Hafele & Keating experiment was clearly *not* an illusion!

In my many discussions with "mainstreamers" no one has come up with a valid reason to dismiss these ideas and the best argument put forward when pressed hard to disprove them was:

> *"This is not accepted physics - I can see where you are coming from, but you cannot average in this way. If you calculate correctly (i.e. a time-weighted sum) then the clocks tick the same but the time is different.*
>
> *The red-shifting/slow-ticking is an artefact of the relative velocity - the time dilation is also an artefact of this; the red-shifting/slow*

> *ticking is not solely an artefact of the time dilation. This explains all".*

This response seems to make the point that *red shift* and time dilation are both perceived as the result of the relative velocity and are due solely to Doppler Shift effects. But, the response also stated:

> "..........*the red shifting/slow ticking is not* **solely** *an artefact of the time dilation"*

This seems to be a subconscious admission that it is at least partly due to the time dilation and this is the point I am making. If you take the classic Doppler effects away, then you are indeed left with the *red shift* (and slow ticking) from the inertial time dilation. This is what the mainstream argument omits, the "*Transverse* Doppler Shift[43]" which is nothing to do with Doppler effects. In fact, it would be better named simply "Transverse Shift" without the word "Doppler". Transverse shift is purely due to time dilation and it is the *red shift* observed at the instant two clocks pass each other, without any closing speed or departing speed, i.e. without classic Doppler shifts. It is the *red shift* or time dilation due to the difference in time rates predicted by SR, verified by Hafele & Keating and quantified by the Lorentz Factor defined later in this chapter.

Another issue is one of cause and effect. The mainstream offers no mechanism or cause and effect for inertial time dilation and today, this remains an unsolved mystery, which no one, it seems, has been keen to address. The new theory now resolves this long-neglected problem and later in the book the physical cause and effect of inertial time dilation is revealed, together with other important, as yet unanswered, questions.

Clearly, I am at loggerheads with the mainstream and the only way to prove either argument is to carry out my proposed experiment, outlined later, to measure the red and/or *blue shifts*

---

[43] Transverse Doppler Shift – Shift of light wavelength due to the difference in time rates between observed and observer (Not classic Doppler shift due to relative motion).

observed from *both* frames in relative motion (accelerated to different degrees).

Mass induced time dilation

In addition to inertial time dilation, General Relativity demonstrates that the presence of mass dilates time in its vicinity, compared to a region of space-time remote from (and unaffected by) any such bodies and their gravitational fields. Extreme examples of mass induced time dilation are black holes. As stars reach the end of their lives and they run out of their explosive fuels, they start to collapse under their own gravity. The outward pressure reduces from the dwindling nuclear fusion. As they collapse and their density increases, their gravitational field strengthens and, if their initial mass was large enough, they eventually compress to such a density that their escape velocity increases to the speed of light. At that precise moment nothing, not even light itself, having the fastest speed attainable, can ever escape from them, thus the term "black hole". The surface of the collapsing sphere at which the escape velocity becomes light speed is known as the "event horizon" of the black hole. The radius of this sphere is called the Schwarzschild radius.

There is recent speculation (Stephen Hawking in 2014), as to the nature of black holes and it has been suggested that, in fact, there are none, but only "grey" holes, where all "additional" matter/energy is trapped in a thin shell between the classical event horizon and a virtual horizon. This negates the everlasting property of black holes and renders their lives finite. This new understanding has been triggered by quantum mechanical theory which does not allow the loss of any information, but I would suggest we need simply to look at the nature of time to come to this same conclusion. If time stops relative to us on the initial formation of the event horizon, then no further contraction is possible relative to the outside universe. All the additional mass/energy falling in later must therefore become trapped above this horizon, in a thin shell. The inner diameter is the initial event horizon and the outer diameter increases as new matter/energy is absorbed by the black hole.

For the classical black hole, at the Schwarzschild radius, time stops relative to our universe. It is, thereafter, a frozen entity for all time. From our perspective, it never shrinks any further. It can grow in size, however, by swallowing up any mass that comes too close and which falls in. We witness this phenomenon in distant black holes, Blazars[44], which feed on mass and emit radiation from their poles as a consequence.

At this event horizon of a classical black hole, an image of someone falling in might be perceived by a remote onlooker to be frozen in time (since time stands still on the event horizon relative to the rest of the universe). It is indeed the horizon of events and, beyond it, nothing can be detected from outside the black hole. In practice, the onlooker will not even see this last image, frozen in time. The light from the image will never reach his retinas and the unfortunate victim, falling into the black hole, will appear to have red-shifted and dimmed out of sight in the blackness by the very moment his image freezes.

We can understand that on the event horizon of a black hole, time stops, and will remain at a standstill for all eternity, (well almost).

This "permanency" is because no radiation, matter or any effect or consequence of events can ever reach back into the universe that created the black hole in the first place. This is the case, whatever happens beyond the event horizon and within the black hole during any hypothetical, further collapse of the star. However, since time has stopped at the event horizon, then I assert that no further contraction *can* happen.

To make the Earth into a black hole, it would have to be crushed down to a diameter of just 9mm, or the size of a peanut. For our Sun, the event horizon would be only about 1km diameter. Clearly, there is no force in the universe that could do this, but if a star is ten to fifteen times the mass of our Sun or greater, then its gravitational attraction, in the absence of explosive nuclear reactions, is sufficient to create a black hole.

An analogy for a black hole is often described, somewhat inaccurately, as a permanent "dent" in the "fabric of space-time". More realistically, I suggest, would be to describe it as a spherical region of space permanently isolated from our universe by the

---

[44] Blazar – A massive, "feeding" black hole that emits regular bursts of radiation of a few minutes duration from the poles, due to material falling in to the black hole.

halting of the passage of time. If you stop time, nothing can ever happen from that frame of reference to affect our universe. All outward communication has ceased and the effect has therefore become permanent. This total isolation is mathematically derived from GR and is now supported by observation, although the new ideas about grey holes and the slight energy leakage (Hawking Radiation) cast doubts about their permanency.

Clearly, black holes are the extreme case, but this dilation of time is present to some extent with all mass. We can measure the time dilation field around the Earth for the GPS. The time dilation is at its maximum at the Earth's surface and then decays, with increasing distance from the Earth, in accordance with a known formula (which we will use later).

It may also be true that a body falling into a type of black hole, having a very steep gravitational field (tide), would be "spaghettified" because "Space-Time" is distorted to such an extent, over short distances, that the very atomic fabric of the body is torn apart. This is often misinterpreted to mean that space itself has undergone a physical distortion of say, our cubic metre (mentioned earlier), when falling into the black hole. But, our cubic metre cannot be distorted, even within such strong gravitational fields. The distortion is real, but it is the distortion of the passage of time only. It is the extreme time rate gradient across the matter contained in our otherwise unchanged cubic metre that destroys it.

The destruction of physical material by these very strong gravitational "tides", when falling into this specific type of black hole, is simply an extreme case of time dilation. The time gradient here is so steep that the passage of time is at significantly different rates over discrete lengths of the material. It is this differential, between the "rate of existence" across these short lengths, that overcomes the atomic bonds that hold the atoms together, and the material simply drifts apart, like bacon from a slicer.

In everyday life, of course, we deal comfortably with the imperceptible variations in time rate in our immediate surroundings. These tiny variations have no effect on the stuff from which we are made. The time rate gradient which runs through us due to the Earth's gravitational field is very small and is insufficient to affect the bonding forces within our atoms.

Thankfully then, we are not all ripped apart by the very small (but nonetheless real) variations in time rates within our immediate vicinity and throughout our physical structure, when we are in the Earth's time dilatational field.

So, there are two things to bear in mind when forming the basis of our theory, namely that both motion and the presence of mass, dilate or slow down time.

We can have the utmost confidence in these ideas since they are both accepted scientific facts, proven by experiment. Yet, we must bear in mind that in our experience of nature, we invariably find that there is only one cause for any one fundamental effect (such as time dilation).

It will be explained later that there is really only motion induced time dilation since mass induced time dilation is demonstrated to be, fundamentally, motion induced.

Inertial Time Dilation

Using simple geometry, given the constancy of the speed of light, Special Relativity shows that the time rate, as well as the physical dimensions, must reduce for a moving entity, relative to the stationary frame. This theory agrees with these findings, but there are issues with interpreting which effects are real and which are purely relative.

It has been scientifically established that time does vary between accelerated frames (different gravitational potentials and/or relative speeds), even though an occupant in the accelerated frame does not directly detect the change in the time rate from within his own frame of reference. The laws of physics remain unchanged for him, whatever his relative speed. Even his electrons still rotate about his nuclei at the same velocity, *but* only in relation to his slowed clock. The time rate within his frame remains unchanged (as far as he is concerned) despite his "clock" (his rate of existence) slowing relative to other, less accelerated, frames.

The experimental confirmation of inertial time dilation in a moving frame is made from the lost time on the moving clock, after it has been in the accelerated frame (Hafele & Keating.) This observation is made in the stationary frame after the clock has stopped moving. We note that the clock has lost time due to its previous motion. So, the geometric proof from Special Relativity that speed slows time is indeed correct in both the relative and the real sense. Time dilation due to motion is *real* and because of its reality it will be a *cause* for certain *effects.* Length contractions or distortions of the vacuum can *never* be a cause for *any* effect since they are merely relative illusions or mathematical abstractions. They are the *effects* from the time dilation and so they are not available as causes for other effects. Only time can be a cause.

Regarding the *physical* changes from being in an accelerated frame, it is not possible to find similar proof that length contraction[45], for instance, is real, since an object returning to the stationary frame bears no evidence of such an effect.

In the accelerated frame, there are apparent contractions to the physical dimensions in the direction of travel (as viewed from the stationary frame), but no such changes are apparent within the moving frame to the dimensions (or even to the time rate, which we know *does* change). We can see that, despite the change in time rate being "invisible" to the traveller, he should nevertheless be able to detect any real changes in dimensions from the proportions within his moving frame, yet he does not.

His proper lengths remain unchanged, so the length contraction predicted by SR is unreal. It is purely relative.

Special Relativity is contradictory since it proposes length contraction, yet it also states that proper lengths remain unchanged. Length contraction is simply the observed, relative effects due to the constancy of the speed of light. Remember,

> *Only real events can have a real effect in a different frame and in fact, real events MUST have a real effect in different frames.*

From this, we deduce that there is only one possible cause for any and every effect and that is the inertial time dilation from Special Relativity. Effects or events observed from a different frame may or may not be real. They may be relative observations of events

---

[45] Length contraction – Special Relativity proposes that lengths contract in the direction of motion as a consequence of the motion.

which are possibly none existent in the frame of the event. We should therefore mistrust geometrically derived predictions of relative "observations" unless and until they are proven by experiment. The following arguments are dealt with rigorously in the next section, *Time Dilation Proof,* but a general explanation is given here by way of an introduction.

We are sure from our mathematical proofs from thought experiments and also from real experiments (H&K) that time is slowed down in the moving frame and that this time dilation is *real.* It is a relative observation from another frame. It is also a "real" effect, within the moving frame, since it has a permanent effect when returning to the "stationary" frame – or indeed on entering any other frame (your clock *was* running slower and so you lost time, on your return, as a result).

The length shortening is considered by the new theory to be merely relative and unreal. We reach this conclusion from the physical null effect when entering another frame. The traveller has not been distorted in any way. We also make the same deduction, indirectly, from the reality of the time dilation. From this reality we conclude that any time related occurrence must be affected in the moving frame by the time dilation in that frame.

This is where we disagree with accepted interpretation of Special Relativity. The velocity in the moving frame, being a time related occurrence, must have increased because of the time dilation and without any force being applied. Accepting this argument, we do not have to hold this velocity equal in both frames. As a result, we do not need to assume the existence of length contraction in order to protect the mistaken assumption of an equal velocity in both frames. As we shall see in a moment, the geometry dictates we have either length contraction or velocity increase, but not both. Mainstream SR cannot make up its mind and chooses both time dilation *and* length contraction. I choose acceleration to be the resulting real *effect* from the real time dilation, the *cause.* I ignore length contraction since this is a purely relative observation and not real. Length contraction is a matter of opinion and is not mathematically imperative.

Traditional Special Relativity considers it essential that velocity must be the same as observed from both frames. The proposed theory defines only the distance travelled to be the same in both frames and this is verified at the end of the journey, (or indeed at any stage of the journey), when both the stationary observer and the traveller *eventually* agree on the distance covered. What they *never* agree on is the time taken to cover that distance and they therefore disagree on the *velocity*.

A change in time rate will affect anything having a time component, like velocity, but purely physical dimensions or indeed any scalar quantity, have no time rate attribute and so *cannot* be affected by speed and the associated time dilation.

We will now see why this new way of looking at things makes no practical difference to predictions from accepted science and, hence, why this has not previously been identified as an issue.

If we consider a moving electric field and a moving plate being the emitter of this field then, from traditional Special Relativity, the moving plate is assumed to undergo physical length contraction, in the direction of motion. This justifies the increase in intensity of the field, caused by the compressing of the field energy into the apparent shorter length as observed from the stationary frame. The new theory uses the increased velocity, as opposed to the length shortening to predict the same, associated, field strength increase. The answers from both assumptions are identical, since the increase in intensity of a field emitted from a shortened length L / γ moving at velocity v, is the same as that of a field emitted from a non-shortened length L moving at increased velocity, $v\gamma$. The following demonstrates this argument in simple mathematical terms:

$$\text{Field Energy} \propto \text{flux per unit area per second}$$

$$\text{Field Energy} \propto \text{Flux} \times \frac{\gamma}{L} \times v \propto \frac{Fv\gamma}{L}$$

Or,

$$\text{Field Energy} \propto \text{Flux} \times \frac{1}{L} \times v\gamma \propto \frac{Fv\gamma}{L}$$

The energy of the moving field has to be emitted and received in full, so, it has to be done in a shorter time and from an apparent, relatively shorter, passing length, as received in the stationary frame. To an observer in the stationary frame, it is as if the length is shortened, but this is just a relative effect and does not reflect the reality in the moving frame. The reality, in the moving frame, is that the proper length is maintained, but time is slowed down and velocity is increased, as a result (velocity of the stationary frame, relative to the moving plate). The plate moves past quicker and the length looks shorter, since it takes less time to pass. This real velocity increase is the cause of the compression of the field received in the stationary frame. We do not have to accept that a purely relative, unreal occurrence, such as length contraction can have a definite real effect in a different frame. From our definition of reality, this can never happen. Mainstream science has taken the length contraction route, but the above example demonstrates the answers will be the same, whichever choice we make, the right one, or the wrong one!

Looking at this issue from a "higher level", we have already demonstrated that the only entity in the vacuum is the time rate field and therefore the time rate is the only available cause to affect events. Clearly, there is no available mechanism to affect the non-time-related physical dimensions unless we introduce the idea of the distortion of space. Accepting this reasoning drives us to conclude that only time-related occurrences (e.g. velocities) are affected by the time dilation of motion and that all physical properties (length and mass) remain unaffected within the moving frame. This is demonstrated to be the case on re-entering the stationary frame. This basic flaw in the reasoning of Special Relativity is due to a lack of understanding of the fundamental difference between time and the physical dimensions. Relativity treats all dimensions the same whereas, in fact, they are not.

Relativity is selective about the reality of events that are relative. Indeed. it seems to abandon the concept of reality altogether in favour of pure relativity. Conventional relativity is an *observational* theory and deals with relative observations between different frames of reference. The new theory is a theory of *realities* (within any frame) and is based on physical and temporal, *cause and effect*. A more rigorous analysis follows, clearly demonstrating this fundamental error of choice between the options of holding either velocity or length constant across different frames in relative motion.

Inertial Time Dilation - Proof

The detailed explanation of inertial time dilation from Special Relativity is easy to understand and the following is the standard proof, accepted by current science. I have chosen the simplest possible example.

Figure 2 shows a static light clock on the left, of such a height that a pulse of light takes one second to travel from the light source, at the bottom of its tube, to the mirror at the top. There is another mirror at the bottom of the tube and so the clock then "ticks" every second. To make the period one second, the tube has to be very long (approximately 300 million metres). The dimensions could be scaled down in proportion to make a more practical thought experiment but for our theoretical purposes, we will leave things as they are, as this makes the numbers very simple indeed.

Another identical, synchronized clock moves past the stationary clock as both clocks tick simultaneously with the light pulses bouncing off the bottom mirrors. It moves in a direction to the right with a rate of "v" units of distance per second of the stationary clock.

After one second has passed in the stationary frame, the light ray in the *stationary* clock has reached the mirror at the top and the moving clock has reached position "v" on the horizontal axis at the bottom of the diagram. Also, after one second has passed in the stationary frame, the light ray in the *moving* clock has travelled the real angular path shown of length "c" (the angle has been exaggerated). This is because the relative speed of light is a constant in all reference frames. The speed of light is invariant.

As a result of this constancy, the light ray has *not* quite reached the top of the moving clock after one second has passed in the stationary frame. The moving clock will have to reach position "$v\gamma$" for the light to reach the top of the moving clock at time "$c\gamma$" in the stationary frame. But, the time shown on the moving clock at this event will be only "c" since the light will have just reached the top of the moving clock. The time rate (clock) in the moving frame therefore appears slow, to an observer in the stationary frame.

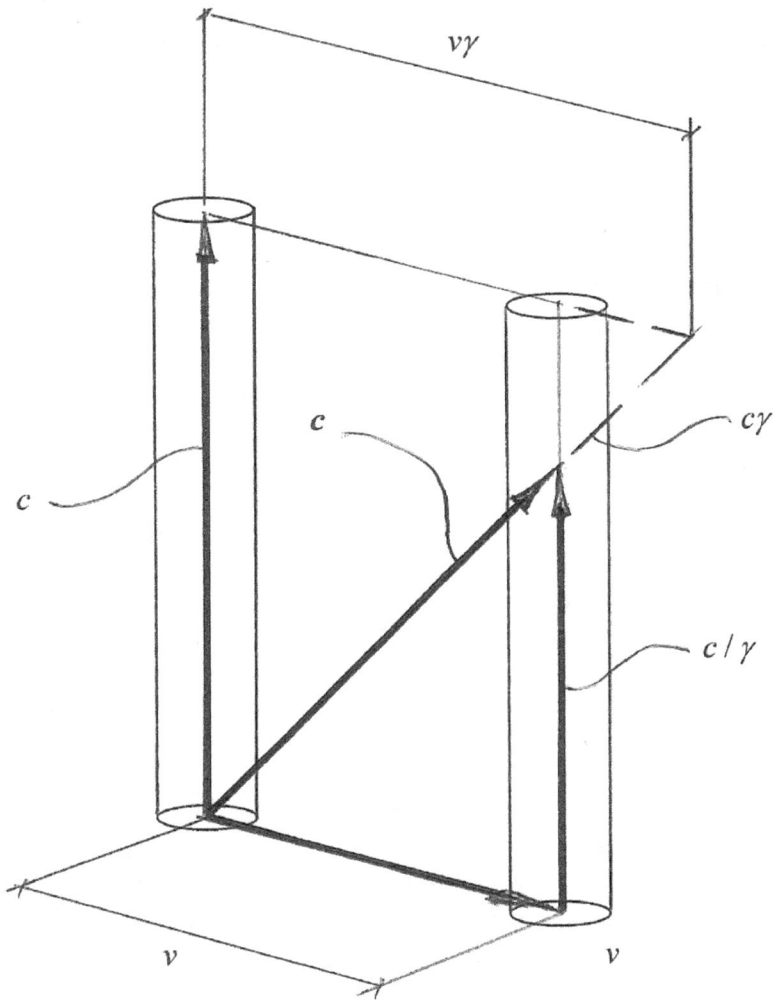

Figure 2: Inertial time dilation and length contraction

This is the standard explanation for inertial time dilation since the moving clock has taken longer than the identical stationary clock by a factor of γ *(gamma)*, to tick once.

It is important to understand *why* this has happened. In this thought experiment, we have fed into the logic, the constancy of the speed of light. If we had not done this, then we would not have derived any inertial time dilation. It is also worthy of note that the geometry we have just carried out, although demonstrating the inevitability of inertial time dilation, does not provide any causality. The geometry, coupled with the invariance of the speed of light, merely demonstrates that time dilation will always occur with linear motion, but it does not demonstrate cause and effect. The constancy of the speed of light is the clue as to the physical cause and we shall explore this, in more detail, later and uncover the causality of inertial time dilation. We shall see in a moment that "γ" has values between one and infinity, but that, for speeds much less than "c", it practically equals one. It is not until we reach speeds of a significant fraction of "c", that "γ" becomes significantly more than one. So, in practice, we do not normally notice this tiny effect, often measured in fractions of a microsecond, say for a jet airliner over a flight of many hours.

The proposed theory is in full agreement with all of the above, but this is just a thought experiment. We must remember that this effect of motion slowing down time is *real* as well as relative and that this has been proven by experiment (Hafele & Keating - 1971).

When viewed from the stationary frame, the distance covered after one second is "v". Mainstream Special Relativity next *chooses to believe* (it is not mathematically imperative) that the distance covered, as viewed from the moving frame and after one second has passed in the moving frame, also has to be "v" in order to maintain the same velocity "v" per second, in both frames. So, velocity is held equal between the two frames and the consequence of this is a length contraction in the moving frame and in the direction of motion, from "$v\gamma$" to "v". The geometry alone dictates this, but only because we have made it that way. The length contraction is the inevitable result of the velocity being held equal in both frames and it is necessary in order to resolve the different distances covered in each frame in the same time as measured by both clocks.

Quite simply, this does not make sense and, on this point, we must disagree with the conventional understanding of Special Relativity.

Relative velocities *cannot* be the same in each frame, when the time rates are different. We *cannot* conclude that lengths in the moving frame have contracted in the direction of motion, simply to maintain the same constant velocity in both frames. There is no mechanism to produce this length contraction. There is no cause for this effect. Length contraction is a purely relative observation, a distortion of reality and is not real.

Velocity is the fraction, distance/time and if you change the time units, you change the velocity. Since the time units are different between both frames then so must the velocities be different as a result. The Principle of Relativity requires that the laws of nature in the moving frame cannot be affected by the frame's constant motion and the laws of nature include the maintenance of physical dimensions. Proper lengths cannot be affected by motion. Special Relativity also agrees with this, dictates even, but then seems to throw this rule away and engages length contraction in order to reconcile the real effects of time dilation. The mainstream interpretation of Special Relativity, regarding length contraction, breaks its own Principle of Relativity.

We know that time is dilated in the moving frame and so, *inevitably,* the constant velocity "v" observed from the stationary frame will be greater when observed from the moving frame, simply due to the relatively slowed time rate. Under these circumstances, there is indeed a mechanism for producing this effect, velocity is a time related occurrence or event and it will be affected by any change in the time rate. Length, on the other hand, is *not* a time related occurrence and cannot be affected by a change in the time rate. We **must** conclude that the "proper velocity" observed from the moving frame really has increased to "$v\gamma$".

This conclusion is deduced from the fact that time has dilated in the moving frame by a factor of "$1/\gamma$", so the distance covered, per dilated second in the moving frame, is now greater, by a factor of "γ", than that covered for each un-dilated second in the stationary frame. The "proper" velocities are different purely due to the different time rates.

This is not to be misinterpreted as meaning that the velocity "v" relative to, or as viewed from the stationary frame, has increased. It has not. It is more that, in the moving frame, everything speeding past is now doing so with velocity "$v\gamma$". It is the velocity of the stationary frame, relative to the moving frame, that has increased by a factor of gamma, because time now passes relatively faster in the stationary frame. The velocity, as observed from the stationary frame, is not amplified by the time dilation of motion. It merely increases in accordance with Newton's 2nd law as we accelerate towards the speed of light, "c". This maximum speed limit is imposed by the time having been dilated to zero and the consequential infinite velocity in the moving frame. This is why "c" is the universal speed limit.

It is what has happened to the passage of time, within the moving frame, that makes any further acceleration impossible.

Perhaps the most important aspect of this disagreement with the mainstream, is to do with the relative motion. The accepted idea, in Special Relativity, is that clocks in the stationary frame will *also* appear slow to the *moving* observer. The value of relative velocity is considered by SR to be the same as viewed from either frame, since velocity is understood as purely relative and it does not matter which frame has accelerated. This view not only matches the apparent length contraction, when viewing the other frame, but it is also necessary to sustain the belief that there is no preferred reference frame, such as the Aether. This is yet another disagreement with the mainstream and we will establish, later, that there is indeed a preferred reference frame or field.

Another reason why velocity is declared equal in each frame, is to avoid faster than light travel, but there is no need for concern over this. The speed limit relative to any observer is still maintained at "c" and "c" remains invariant. It is simply that, from the moving entity's point of view, it starts to travel faster than light as soon as the relative speed exceeds a value where the sum of the relative speed plus the acceleration due to time dilation becomes equal to "c". From here on in, it becomes increasingly superluminal within its own frame. This value of relative velocity, where the proper speed becomes equal to "c" is given by:

$$v = \frac{c}{\sqrt{2}} = 212 \times 10^6 \, m/s$$

Of course, this is a meaningless number, since the universe is not really moving at light speed relative to the traveller, in the conventional sense. Yes, the numerical value of velocity relative to your spacecraft is 300 million metres per second, but this is without the associated time dilation relative to you. In fact, the clock of the universe is fast compared to yours, so there is no limit to the speed of the universe relative to you, until your clock stops altogether.

I quote here a statement made by Dr. Lawrence Krauss, in one of his regular radio interviews, as he refers to the photon experiencing a zero time rate:

> *"A photon sees the complete life of the universe until the end of time, in an instant"*
> *(Unless or until it hits something)*

> *Dr. Lawrence Krauss*

> *ASU Radio interview*

This is a clear admission that the observation from the moving frame is *blue shifted*, contrary to the traditional viewpoint of SR. It seems Dr. Krauss and I completely agree on this matter.

The traditional line of argument within Special Relativity is clearly flawed, since it has been proven in 1971 by Hafele & Keating that it *does* matter which frame has accelerated. The accelerated clock really did run slower than the one that did not accelerate and it is nonsensical to argue that the slowing down of time, in the moving frame, is purely relative and is not real. The stationary clock did not run slow! Wilfully, the scientific community has ignored this decisive result in order to hang on to the notion of velocity being purely relative, despite the indisputable experimental evidence to the contrary and all in the attempt to engineer the avoidance of the dreaded, preferred reference frame.

The opinion that all motion is purely relative, *cannot* be true, since there is a limit to relative speed, the speed of light. If motion were purely relative, there would be no limit and we could go on accelerating relatively, indefinitely. Our speed would be only relative, after all. The only meaning we can attribute to this limit, an apparently arbitrary value of 300 million m/s, is that time stops in the moving frame on attaining this speed. We conclude from this that it is the stopping of the moving clock that is the cause for the inability to move any faster. To date, we have simply accepted this universal speed limit without determining why, without establishing the cause and effect. The causality for this will be revealed later.

The new theory proposes that the stationary clock will actually appear fast, or *blue shifted,* when viewed from the moving frame and this is consistent with the amplified velocity experienced in the moving frame, as explained above. The time rate in the stationary frame has not changed, since there has been no changing of frames, unlike the moving entity, which has accelerated from the stationary frame and does have a reduced time rate as a result. This is an experimentally proven effect and this effect is *definitely* real. As previously mentioned, this effect is known as the "Transverse Doppler Shift", although it is not really a Doppler effect, but is due to the slower clock, in the more accelerated frame, the "moving" frame.

Classic Doppler effects apart, when you look into the moving frame, everything appears *red shifted* since the clock is slower relative to the stationary frame. This effect is in addition to the classic Doppler effects caused by the relative motion. Since this moving clock is slower, then the stationary clock must be relatively faster (again ignoring the classic Doppler effects). We can deduce that, when you look into the stationary frame from the moving one, we will observe a *blue shift,* with everything appearing to happen faster within the stationary frame, including the relative velocity. In the case of the photon, it all happens so fast that it sees the end of time in an instant, according to Dr. Krauss.

If you have difficulty in accepting that the *red shift* of time dilation is not reciprocal, then bear this in mind – Hafele & Keating proved, in 1971, that clocks run slower, the faster you travel. This is a *real* phenomenon and it is not just clocks, but time itself that passes at a slower rate in the moving frame, relative to the less accelerated frame. This means that all events in the stationary frame occur relatively faster than they do in the moving frame. If we consider the event of speed, then speed in the less accelerated frame will appear faster when viewed from the more accelerated frame. That means the *same* speed, viewed from the different frames, will be different from each frame, the complete opposite to the mainstream's perverse assumption of a common velocity viewed from all frames.

It would be fairly simple to perform an experiment that demonstrates this effect and this is detailed later in the book. The experiment must use two light sources of the same, known frequency, one having been accelerated to great speed, the other, less so. The predicted transverse shift will be a result of the difference between the acceleration of each light source from the moment they were initially "synchronised". The frequency measurements must be taken at the instant the two light sources pass each other to minimise classic Doppler effects, although these could be calculated and removed from the results.

Clearly, the timing of the measurement needs to be adjusted to compensate for the travel time of the light between sources/detectors, but the experiment would be best served by a very close passing distance to minimise this, in effect, a deliberate "near miss".

The relative observation of a slow clock in the stationary frame, when viewed from the moving frame, has not yet been experimentally verified. For example, Muons travelling close to the speed of light seem to slow down their decay, when travelling through the Earth's atmosphere, because of the *red shift* in the moving frame.

But, so far at least, no one has managed to sit on a Muon and confirm the predicted *red shift* of the experimenter!

The idea from SR that you will observe *red shift,* when looking into the stationary frame, remains speculative. It is based on opinion and theoretical geometry, not on observation. It takes no account of the real time dilation and its asymmetrical effects.

Actual pictures of muons have been taken showing them to be egg-shaped due to length contraction. They are slightly flattened in the direction of motion. There is no dispute about relative observations. Indeed, lengths will *appear* contracted from the stationary frame but, again, the proper diameter of the muon must remain constant in the frame of the muon and this is predicted by SR itself.

The geometry which tells us that each frame will appear to have a slow clock when viewed from the other is incorrect and does not reflect reality. It tells us this, because the geometry is based on physical dimensions which are symmetrical or bidirectional, whereas the phenomenon of time is not. Time is asymmetrical or unidirectional. Since space, on the macro scale, is effectively a void and time is the only real entity, then the time dilation is the only reality and observation predictions must be made based on the time rate, not just the physical geometry.

We *must* satisfy the temporal aspects even if we have to override the geometric ones.

In relativity the clock is *always* real but the geometry is ultimately just lines on a piece of paper.

Acceleration Due to Time Dilation

Inertial time dilation can be envisioned as the reduction of the *dt* term in the simple "simultaneous" velocity equations:

Velocity in the stationary frame $v_s = \dfrac{dl}{dt_s}$

Velocity in the moving frame $v_m = \dfrac{dl}{dt_m}$

(Velocity = distance/time, or increment of distance *dl* divided by the corresponding increment of time *dt*)

Note, we are holding length constant and allowing only time to vary between frames. Even though we have lost simultaneity between frames due to the time dilation, this is not important. Velocities are constant in both frames, even if they are different, so it does not matter at which precise moment we choose to observe either frame, we will see the same speeds at any moment over time.

Because neither frame is stationary in absolute terms, then for each second on a remote, fast ticking clock t_f in a theoretical "absolutely" stationary frame, the observation is that dt_s is shorter, but that dt_m is shorter *still*, due to the time dilation caused by velocity "v_s". So, dt_s is longer or larger than dt_m.

Therefore,

$$\frac{dl}{dt_m} \geq \frac{dl}{dt_s}$$

And,

$$v_m \geq v_s$$

And so, from our thought experiment,

$$v_m = \gamma v_s$$

This tells us that the velocity, relative to the stationary frame (but as observed from the moving frame), is greater (*blue shifted*) by a factor of gamma than the velocity relative to the stationary frame and observed within the stationary frame. By the previous definition of reality, we see that this increased velocity in the moving frame is *real,* since it is observed from its own frame and has a permanent effect when re-entering the stationary frame. This permanent effect is that the distance travelled has been achieved in a shorter time for the traveller, than the time measured for the same journey in the stationary frame, just like the muon. Ultimately, the velocity in the moving frame equals the total distance covered as we come to a halt, divided by the time taken in the moving frame to cover that distance. This is the *real* or proper velocity. The relative velocity is merely that, relative.

We have additional speed in the moving frame, simply because it is moving in the first place. The time dilation and its associated acceleration are progressively induced as the acceleration from the stationary frame is applied.

To be clear, if the acceleration is applied to give a constant velocity v relative to the stationary frame, then the velocity in the moving frame, relative to the stationary frame, is greater due to time dilation and gives a constant velocity in that frame of $v\gamma$. From this moment onwards, assuming no further acceleration, both frames will have constant velocities of v and $v\gamma$ respectively. We will now present this in a more mathematical way:

Firstly, γ is defined by using Pythagoras for the small triangle c, v, c/γ, in Figure 2. And so,

$$\gamma = \frac{1}{\sqrt{1 - v^2/c^2}}$$ The Lorentz Factor.

(Where v is v_s, the velocity relative to and as observed from the stationary frame)

Therefore,

$$v_m = \gamma v_s = \frac{v_s}{\sqrt{1 - v_s^2/c^2}}$$

We can see from this equation that as v_s (the velocity relative to and observed from the stationary frame) approaches c, then v_m (the velocity experienced within the moving frame), approaches infinite speed, since the denominator on the right hand side approaches zero.

So as

$$v_s \rightarrow c$$

Then

$$v_m \rightarrow \infty$$

The argument may arise that we cannot have infinite velocity in the moving frame since this would break the speed limit of "c", the invariance of the speed of light. The new theory argues that we have not exceeded the relative speed limit of "c", but only that the proper velocity experienced in the moving frame, has exceeded "c" and this is purely due to time dilation. No force has had to be applied to produce this additional acceleration which has occurred progressively in accordance with Lorentz, as the relative velocity observed from the stationary frame has been increased to "c". Current science agrees that we can approach velocity "c" relative to the stationary frame and this is proven (by the above) to approach an infinite velocity within the moving frame, since time stands still in that frame at a relative velocity of "c". When your clock has stopped altogether, any speed at all becomes infinite when looking sideways out of your porthole, just like for the photon.

This justification can be further reinforced by considering that the time rate dilating to zero in the moving frame is accepted by current science, is proven by mathematics and is supported by experiment. At a time rate of zero, any velocity at all becomes infinite and you get from A to B instantly. This proves the limit of "c" is a false one when considering the reality in the moving frame and that it only applies to velocity relative to and as observed from other frames. *Within* the moving frame there is no speed limit and as the real or proper velocity approaches infinite speed, $v_m \rightarrow \infty$, the velocity relative to and as viewed from the "stationary" frame approaches the speed of light, $v_s \rightarrow c$. All our observations of moving entities will only ever show maximum relative velocities of "c" and this apparent speed limit is what has deceived us for over a century.

To visualise this we might consider what happens to a single instantaneous image in a beam of light travelling through space over a distance of say, several light years. The speed of this image relative to us is "c" and the image takes several years, by our clock, to get to us, since "c" is a fixed, finite speed. However, this same image emitted several years ago by our clock, has *experienced* no time at all in reaching us because the clock in the moving frame has stood still. This is demonstrated by the image being "frozen" for the whole of its passage, frozen in time since its emission from the source. If we were sitting on this image, we would also have got here in an instant because of our infinite velocity as experienced in the moving frame. The only way to get from A to B instantaneously is to travel at infinite speed. Viewed from the stationary frame the corresponding velocity is of course, only "c".

If we consider this image travelling through space and the fact that it does not change in the slightest over its long journey, then we must realise that its entropy has remained constant. Normally, we should expect such an image, consisting of electromagnetic radiation, to degenerate and diffuse over its travels in accordance with the second law of thermodynamics, but this is never the case with light. The only way to avoid any degeneration of the image is for it not to experience any time passage and this is also always the case with light.

All of the above still fits with standard relativity, in that the speed of light cannot be exceeded relative to an observer in any other frame, but we now have to accept that the velocity *experienced* in the moving frame, (and by any beam of light), has increased to infinite speed by the time a *relative* velocity of "*c*" as observed from the stationary frame, has been reached.

One possible objection to the idea of infinite velocity comes from the consideration of the spherical surface of an expanding field of radiation from an emitting source. If we consider the radial expansion to be infinitely fast, as proposed, then the spreading of the field energy around the surface of the sphere as it expands radially must be proportional to $2\pi r$. But r is already increasing infinitely fast, so the objection arises that we certainly cannot exceed infinite speed in the circumferential direction of expansion. There are two defences against this objection. The first is that time has dilated to zero, in the frame of the surface of the expanding sphere, and therefore *any* distance is covered in no time at all, which includes both r and $2\pi r$. When speed is infinite, it does not matter how far you travel in terms of the time it takes, since the clock has stopped completely and distance or length becomes irrelevant to the traveller. The second defence is that the radiation is not moving around the circumference of the sphere, but is merely "stretching" or "thinning". If you consider any particular point on the sphere, then that point, any point, is always moving radially outwards. It is never moving around the circumference and so the objection disappears.

If we accept this subtly different view of motion-induced time dilation, we do not now have to accept counter-intuitive notions (like length contraction which cannot be physically explained or which should not even have been considered in the absence of any identifiable cause). The acceleration due to time dilation does have an identified cause as we shall see later. It is perfectly logical, understandable and inevitable, it is completely intuitive. This does destroy the myth of the limiting speed of light, but only as viewed from the moving frame and the relative limit of "c" still stands. The speed of light remains invariant.

When we compare the limiting effects from this theory with those from traditional Special Relativity, we see that the new theory gives coincident real limits in the moving frame of an infinite speed, a zero time rate and no physical impact within the moving frame. Special Relativity on the other hand, gives coincident limits of a maximum velocity of "c", a zero time rate and a length contraction to zero in the direction of motion. Mass is also envisaged to become infinite but this will not be evident within the moving frame. Like length contraction, mass increase is a purely relative effect on a physical, non-time-related entity and as such is unreal in the moving frame.

It is clearly more acceptable, from both scientific and logical perspectives, to accept the limits from the new theory as compared to those from traditional SR which are inconsistent amongst themselves. Furthermore, the New Theory uses scientifically conclusive, deductive logic and demonstrates an identified cause and effect, whereas the logic used in the derivation of events in Special Relativity is clearly flawed, has no valid "cause and effect" and requires the acceptance of counter-intuitive effects without any physical justification.

The major point to note is that only time related events are affected by the time dilation of motion, because the only entity in the vacuum is the passage of time. The time rate is, therefore, the only available cause for any real effect. Time rate changes cannot affect physical dimensions, since these are not time related, but they can (and do) affect velocity which *is* time related.

Special Relativity has an internal conflict in that, to begin with, it claims that the clock in the other frame will always be observed as slow, whichever frame you are observing from, moving or stationary. On the other hand, it has to admit that the moving clock is the slow one and the stationary clock remains faster, since this has been proven by experiment in 1971 by Hafele & Keating. This inconsistency has never been resolved and the scientific community has chosen to ignore the unavoidable implications from the Hafele and Keating experiment and to hang on to the almost religious belief that all motion *must* be purely relative. The proposed theory takes the results from Hafele & Keating literally and recognises this reality of differential time rates. It does not predict equality of observations between frames, but actually predicts the observation of a *blue shifted*, faster clock in the stationary frame when viewed from the moving one.

In other words:

> *Time dilation is not symmetrical, and so it is never reciprocal*

This is a fundamental disagreement with the mainstream interpretation of Special Relativity which claims that all observations between different frames in relative motion are symmetrical. This is considered to be the case, since the relative speeds are the same between each frame, but this view ignores the fact that one frame has increased its speed through the time rate field and the other has not. The stationary field of time is the preferred reference frame and we will prove this later.

If you are still not convinced, then to more clearly understand the disagreement, we need first to consider the original stationary frame with the two entities initially having their particular, common time rate. Then, one entity accelerates and moves relative to the other. It is a proven fact that this newly occupied moving frame is the one with the slower clock. It is, therefore, the identical, initial conditions and the selection of the one entity that changes frames, which defines the relative time rates during relative motion. The entity that does not accelerate and switch frames does not change its time rate. If we do not arrive at this conclusion of the non-reciprocity of observations, then we break the Principle of Relativity as well as the Equivalence Principle. The Principle of Relativity states that the laws of nature, of physics, are the same in any frames of reference, which move relative to each other with constant velocity.

We must also look at gravitational time dilation, to understand the fundamentals of inertial time dilation. The observed *blue* (and *red) shifts* when looking outward from (or inward to) a gravitational source, are to be more fundamentally understood as the result of looking into a region with a faster (or a slower) clock, respectively. When looking *upwards* from a source of gravitation we are looking into a region having a faster clock than ours and so everything appears *blue shifted*. When we look *into* a source of gravitation, we observe *red shift* since we are now looking into a region with a slower clock than ours. This is accepted science.

Considering this fundamental, temporal effect, but instead for an inertial, time dilated frame, we must conclude that the relative observations must also be the same as for the gravitational equivalent, namely *blue shifted* when looking outward from our moving, *slow* clock to the "stationary" frame with its *faster* clock. In other words, it does not matter *how* the differential time rate is created, either by gravity or by motion or by a combination of the two and we will always observe the relative time differential, the reality.

This not only means that the light is actually *blue shifted* when looking at the stationary frame, and this would be readily observed if we were to carry out my proposed experiment, but also that all events are similarly observed as *blue shifted,* or speeded up. Our observations of velocity will be affected in this way, since velocity is a time related occurrence, an event, and will therefore appear to be faster when viewed from the moving, *red shifted* frame.

At the speed of light, static time really is infinitely faster than our time rate since ours has become zero. This is why we observe an infinitely *blue shifted* static universe and any and all observable, classic Doppler effects are ultimately overpowered by the inertial time dilation at the speed of light. This *blue shifting* is a minute effect at first, since low speeds are insignificant compared to the rate of passage of time. Gradually though (and in accordance with Lorentz), the static universe becomes increasingly *blue shifted* relative to the traveller, due to the inertial time dilation and this really "kicks in" at higher speeds and closer to the speed of light.

Figure 3 shows how the static universe appears to the accelerating traveller to be increasingly *blue shifted,* due to the inertial time dilation, as velocity increases to light speed. Finally, at the speed of light and as the traveller's clock draws to a halt), the static universe appears to evolve infinitely fast, relatively.

Please remember we are referring here only to the *transverse* shift observed through the traveller's "port hole". This effect has nothing to do with classic Doppler shifts.

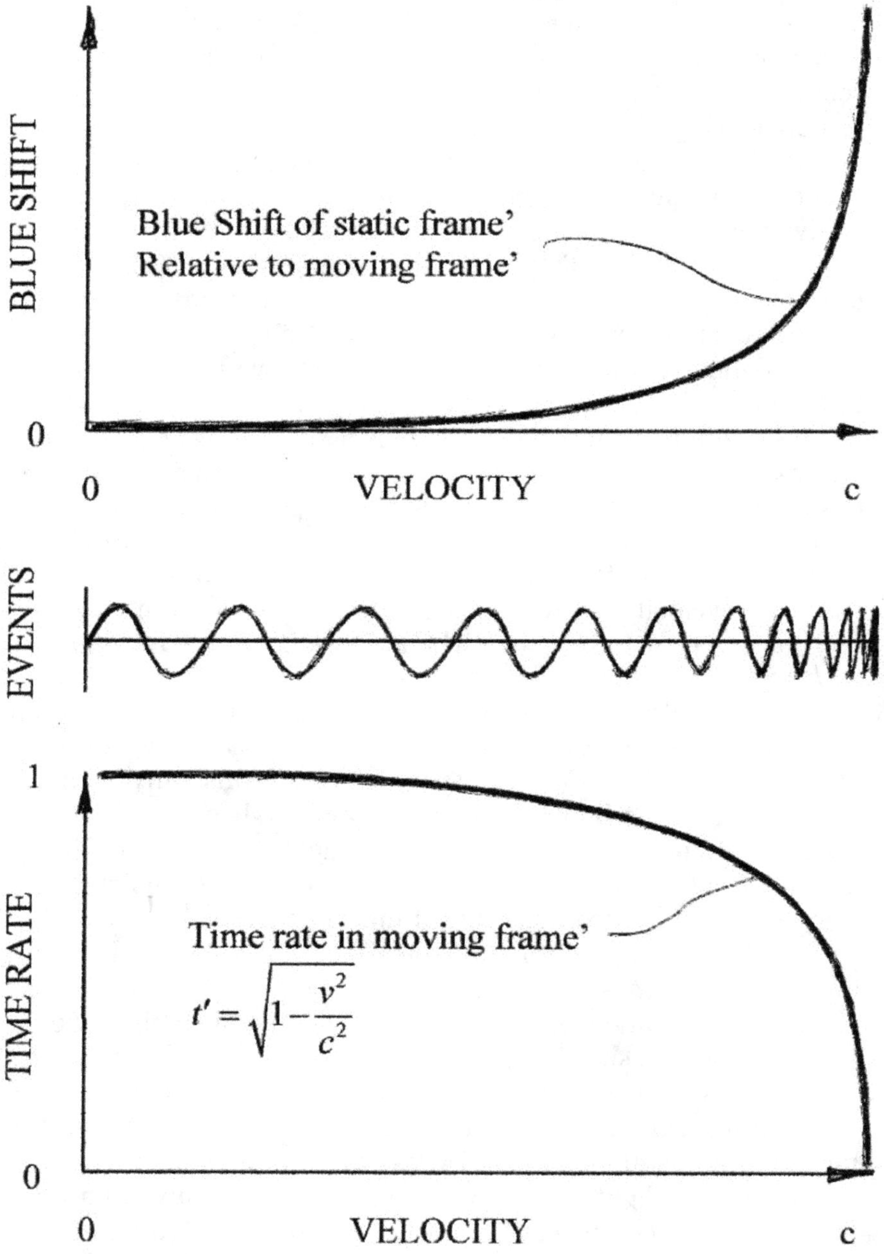
BLUE SHIFT
Blue Shift of static frame'
Relative to moving frame'
0
0
VELOCITY
c
EVENTS
1
TIME RATE
Time rate in moving frame'
$t' = \sqrt{1-\frac{v^2}{c^2}}$
0
0
VELOCITY
c

Figure 3: Transverse *blue shift* of the static universe

The mainstream is confused by Doppler effects which *will* appear red shifted from *both* frames and this is what is currently misunderstood as all motion being purely relative. But, as speed increases toward light speed, any observed Doppler *red shifting* is ultimately overpowered by the real, transverse shift from the time dilation and this is *not* reciprocal.

This is what the mainstream has failed to understand and although it is understood that time dilation is not symmetrical and the "dimension" of time itself is unidirectional, the *consequences* of this are *not* understood. Observing the *moving* frame will show pure *red shift* due to classic Doppler effects, plus an initial, tiny transverse *red shift* due to the inertial time dilation which the mainstream ignores.

The key thing to understand is that this transverse shift is not reciprocal and, although this will show as *red shift* when observing the moving frame, it will show as a *blue shift* of the stationary frame, observed from the moving frame.

Classic Doppler effects *are* reciprocal, so we have been confused into believing that all the frequency shifting is reciprocal, but this is not the case with time. Time is asymmetrical and is therefore *never* reciprocal. Any and all Doppler shifts, *red* or *blue*, are purely relative and so are ultimately overpowered by the real, transverse shift from the time dilation.

The wave form of events in the centre of Figure 3 depicts how much faster the static universe evolves compared to the moving frame with its slower clock. This is not, strictly, a Doppler *blue shift* of any observable, cyclical event, but is indicative of the relative speed of *time* compared to the moving frame. By definition, it is the transverse *blue shift* and it is equal and opposite to the inertial, transverse *red shift* of the moving frame as observed from the static frame.

To press the point home, time dilation is never reciprocal whether it is generated by gravitation or by motion and this is another fundamental disagreement with the traditional interpretation of relativity theory. The mainstream asserts that *inertial* time dilation *is* reciprocal but that *gravitational* time dilation is *not* reciprocal. This is inconsistent, since you cannot have different types of time dilation!

Again, time dilation is a *real* effect caused either by motion or gravitation, so time is either dilated or it is not. The real, relative effects *must* be the same, *however* they are generated. Time dilation is *not* reciprocal for both inertial *and* gravitational systems.

Logically (not merely "intuitively"), because your clock has slowed down, the static universe evolves relatively fast compared to your rate of evolution and ultimately, at light speed, when your clock has stopped altogether, the universe progresses infinitely faster into the future than you do. So, it now takes you no time at all to get anywhere you like, since your speed has become infinite. The penalty on arrival is, of course, lost time and this is determined by how far you have travelled. Inherent within all these arguments are two fundamental disagreements with current, mainstream thinking.

The first and most obvious is this transverse *blue shift* observed from the moving frame and not simply the reciprocal Doppler *red shift* as argued by the mainstream interpretation of Special Relativity. This disagreement hinges on the interpretation of the theory rather than the mathematics. There is no disagreement with the math. Both interpretations are allowed by the mathematics of SR and it is a matter of *opinion* which view you take. Calculations made by the mainstream, using length contraction, do give correct answers, but for the wrong reasons!

If they used acceleration due to time dilation, as opposed to length contraction, they would get the same answers (but at least the calculations would be based on reality rather than an illusion). Nevertheless, the new theory will stand or fall from the results of the proposed experiment to measure the transverse *blue shift* observed *from* the more accelerated frame.

The second disagreement is that there *is* a preferred reference frame or reference field. This is the time rate field and it is movement through this field which causes time dilation, *not* movement relative to any other entity which may or may not be moving through the field. We will pursue this later in the book, but we can ask now,

> *"Why should time be dilated by one's motion relative to any another entity?"*

There can only be one cause for this effect. It *must* be our motion through the field of time which causes time to slow down.

The mainstream has ignored the fact that for every effect, there must be a proximate, antecedent cause. The time rate field is the *only* entity that is proximate to all moving particles including those within gravitational fields.

We have been too clever in our justifications for dismissing the "counter-intuitive" aspects of relativity such as the twin paradox and the idea of all things relative, but I offer, here, a completely intuitive, logical, causally based and physically based interpretation of Special Relativity which uses the same math and gives the same results and so I am claiming that this view is superior to the traditional understanding of SR. It is the correct view and my proposed experiment will confirm this.

To summarise, Doppler shifts are due to the relative motion between the observer and the object emitting a wavelike signal. The wave signal gets compressed or stretched depending on whether the emitter is moving toward or away from the observer. But, the transverse shift is nothing to do with compression or expansion of wavelike, signal transmissions, it is due to the time dilation which also effects the rate of all events and not just the appearance of wavelike signals. The transverse shift will make every event in the static frame happen faster in the eyes of the observer in the moving frame. This is not reciprocated and, although events in the moving frame observed from the static frame will look *red shifted* and slower to a stationary observer, events in the static frame will appear to happen faster from the perspective of the moving frame. This is the key to the misunderstanding. The mainstream has failed to realise the difference between the effects of Doppler shifts and transverse (temporal) shifts.

Twin Paradox

It is worth looking at the resolution of the well-known "Twin paradox", to gain a clear understanding of what the proposed theory means, regarding observations from the moving frame. We will consider two twins, one, a girl, who stays where she is and the other, a boy, who leaves instantly at the speed of light on a trip lasting four light years out and four light years back. The boy will return having hardly aged at all, whereas the girl will be almost eight years older than him. The paradox is that Special Relativity predicts that observations from either frame during the trip will show a slower clock in the other frame. This is in conflict with the experimentally proven time dilation in the moving frame which demonstrates the clocks do indeed operate at different rates. With the proposed theory, there is no paradox, since the predicted observations are different, because of the difference in the real time rates between the two frames. There will, of course, be classic Doppler shift effects with any observations between the two frames, but these are purely relative effects, due to the relative motion, and they do not reflect the reality of the time rate differential. They do not take account of the Transverse Doppler Shift[46].

Since light speed cannot be attained by a body having mass and, since it will take a great deal of time to accelerate and decelerate to and from speeds close to the speed of light, then the example becomes somewhat impractical. It does however demonstrate the principles involved and, in this regard, it remains scientifically sound, despite its impracticalities. We could become side-tracked into carrying out time-weighted sums to nit-pick the exact details but, frankly. this is irrelevant to the points being made here and I prefer to keep things as simple as possible.

The apparent paradox has been addressed by the mainstream, using the arguments that the moving frame has been accelerated and/or that a return journey uses two frames, but these justifications are irrelevant as well as incorrect. The final age difference is purely the result of the time spent by the boy in the temporally "slower", (but moving) frame, whilst the clock ticks faster for the girl in the stationary frame. He has slowed his time down, by virtue of his motion, and allowed the stationary universe to "overtake" him in the temporal sense. The effect has nothing to do with the number of frames used, but is simply due to the total time spent in the other frame(s), the difference(s) in time rates between frames, and the fact that *he* is the one who has changed frames.

This effect can be best understood by visualising the different frames of reference, such as our previously mentioned horizontal

---

[46] Transverse Doppler Shift – shift of light wavelength due to the difference in time rates between the observed and the observer. (Not classic Doppler effects due to relative motion)

travelators (or conveyors) moving at different speeds, but in the same direction. Considering only those that are moving in the same direction, then the movement of the travelators represents the direction of time being always into the future. The different speeds represent the different time rates within different frames of reference. Figure 4 shows a number of such frames or travelators and the space-time paths of the twin boy and the girl are shown. Time passes from left to right and velocity increases positively upwards and negatively downwards from the central "stationary" frame which has the fastest clock.

The figure shows that the boy's speed is *almost* the speed of light, since it is not possible for a physical object to actually achieve this speed but only to approach it and so we have at least kept this aspect of our reasoning within practical limits. However, we have assumed instantaneous acceleration and deceleration, simply to make the point clearly.

The dotted lines, depicting acceleration and deceleration, will be curved, in reality, but this presentation more clearly demonstrates the fundamentals. We can see that the boy first enters a temporally "slower" frame by virtue of his outbound (positive) velocity. Upon reaching his destination, he then does a physical U-turn and enters a frame with the same time rate as his outbound journey, but this is now due to his velocity in the opposite (negative) direction.

We should note that the *direction* of motion has no effect on the time dilation, which is purely due to the numerical value of the relative velocity and not its sense.

Time dilation pays no heed to physical direction, but only to the degree of motion. It is also worth considering that the frame he enters on the return leg is, in fact, the same frame used for his outbound leg, since the time rate is the same and it is the time rate that defines the frame (See Chapter 8, *Some Rules About Time*, Rules 8.1 and 8.9). We can see from the diagram that the boy spends his entire journey in a frame with an almost zero time rate, whilst the girl's clock ticks normally for eight years.

The travelators move (time passes) at different rates as indicated by the number of arrows on each time line. One arrow represents an almost zero time rate in the boy's frame, and the centre time line with the most arrows represents our normal, "undilated" Earth-bound time rate, the girl's frame.

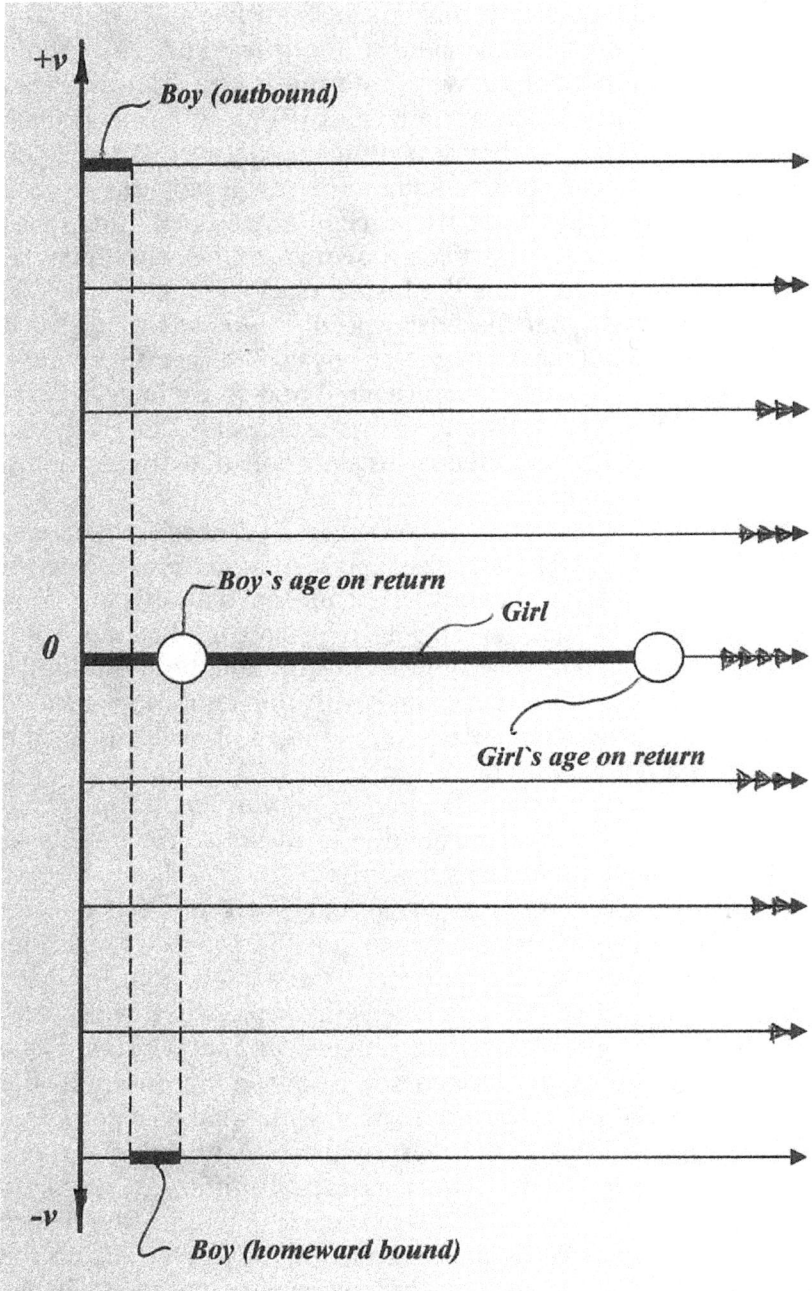

Figure 4: Time Conveyor

We can now see how the speed of the aging process differs between the centre frame, in which time passes relatively quickly, and the moving frame, where time passes relatively very slowly. It is this mechanism of differential time rates that produces the end result of the girl twin being almost eight years older than the boy on his return.

Relativity attempts to reconcile these effects at all stages of the journey. This is indeed important, but this is only a geometric exercise. The above is the fundamental cause and effect for the different ages of each twin on the boy's return.

There is no paradox!

When looking at Figure 4, we are taking up the position of some omnipotent "God" outside of time, who can see everything, everywhere, simultaneously, and we are able to see all the clocks at their different rates.

We do not observe both clocks being slowed down, only the moving one.

Even when we place ourselves in each frame, we observe *red shift* when looking from the stationary frame to the moving one, but *blue shift* when looking from the moving frame to the stationary one. We do not get the observation from either frame, predicted by the mainstream, of both clocks appearing slow when viewed from the other, when we remove Doppler effects.

How do we reconcile this conflict with Special Relativity?

Well, we are identifying the *reality* of the time dilation and taking this to be quite separate from the Doppler effects from the relative motion. We view, separately, the two effects of the Doppler shift and the temporal dilation from the accelerated motion, whereas our observations, in practice, are a combination of both.

Special Relativity, on the other hand, takes the pure geometry as the overriding cause, ignoring the real time rate differential and this is why it deduces the reciprocity of observations. Geometry is bidirectional or symmetrical, whereas the time dilation is not, it is unidirectional or asymmetrical and the reality of time does not recognise physical direction but only the numerical value of motion.

The proposed theory therefore resolves this apparent paradox and also aligns with certain established rules:

- You can only move into the future quicker by slowing down your clock, *not* speeding it up. When your clock slows, then the static universe's clock runs faster and so more time has passed when you re-join it.

- You can increase the rate at which you move into the future by increasing your speed or by staying in a lower gravitational potential. Both of these actions slow your clock.

- You can never move into the past.

5 Newtonian gravitation

The Language of Science - Mathematics

Mathematics and physics take fundamentally different approaches to describing nature. The former is more concerned with what might be possible, and the latter with what is definitely real. Math is constrained by the need for internal consistency, but is generally oblivious to external constraints. Physics has its laws too, and these can change as knowledge improves. But physics is rigorously constrained by its principles which have no counterparts in mathematics. Examples are, the causality principle ("Every effect must have a proximate antecedent cause"), and the prohibition against creation. Violations of such principles are ruled out by logic as requiring magic, a miracle, or the supernatural. Although mathematically allowed, they are said to be physically impossible.

Tom Van Flandern & J. P. Vigier

(Foundations of Physics (32:1031-1068, 2002)

Since the time of Newton, the language of science has been mathematics, even more so since the beginning of the twentieth century. This has proved invaluable for the rapid scientific progress that has been made, particularly in the last one hundred years. Mathematics can give the certainty of proof to theories which may otherwise be in perpetual doubt. It can even make progress on its own, without the necessity for us to visualise the theories, rules or outcomes that it creates, supports and predicts. It can even make discoveries on our behalf and we are then simply left with the task of interpreting what the math tells us.

In these days of the mathematical physicist, this very strength of mathematics, in effectively having its own mind, is a potentially dangerous attribute. It can mean that we, as intelligent beings, perhaps do not, will not, or simply cannot apply our powers of visualising to what we discover mathematically about our universe. We may fail to challenge the mathematical implications or even to get the right visualisation and initial application of the math right in the first place.

> *"Today's scientists have substituted mathematics for experiments and they wander off through equation after equation and eventually build a structure which has no relation to reality."*
>
> *Nikola Tesla (1857 – 1943)*

We might create mathematical four-dimensional space-time models for instance, to try and visualise what the mathematics of our space-time continuum means. These models would be very limited in their ability to help us achieve these visualisations. It is also possible that, when we use mathematics to take the lead in expanding our understanding, we simply get left behind conceptually and consequently overlook some fundamentals.

It is arrogant to believe that human conceptualisation will no longer be valuable in our search for a more thorough knowledge of our universe, that mathematics is the panacea for scientific progress.

This attitude derives from an over-reliance upon mathematics and we seem to be losing our ability to improve our understandings using visualisation and deductive logic.

These days, every physicist must be a mathematician, but being a good mathematician does not necessarily make a good physicist. It is also worth remembering how creative we can become, when faced with things we do not fully understand. Curved space, I believe, is one of these challenging creations.

The notion of curved space-time and of curved space (General Relativity (1915)) was presented fairly soon after the idea of flat space with curved time (Special Relativity (1905)). It may be that the later idea was too eagerly accepted, before we had fully tested how the curvature of time behaves on its own. This could be considered a fundamental breach of the scientific method, in that, effectively, we had altered two variables almost at once, instead of just one at a time. As a result, we seem to have lost the opportunity in the early part of the twentieth century, to completely explore how gravity could work in flat space time.

Scientists in relativity agree that Newtonian gravitation can be viewed as the curvature of time in an otherwise locally flat space-time. Minkowski space-time or flat space-time was proposed in conjunction with and as a result of, Special Relativity, but apparently it does not quite fit with the later theory of General Relativity. General Relativity does include the notion of flat, Minkowski space, but only as a special case and not the general one.

A general theory requires that it can cope with *any* model of space-time, flat or otherwise, and the issue then becomes one of identifying how, why, where, when, and of course *if*, space itself might become curved.

> *Locally flat space time:*
>
> *Strictly speaking, the use of Minkowski space, to describe physical systems over finite distances applies only in the Newtonian limit of systems without significant gravitation. In the case of significant gravitation, space-time becomes curved and one must abandon Special Relativity in favour of the full theory of general relativity.*
>
> *Nevertheless, even in such cases, Minkowski space is still a good description of an infinitesimally small region surrounding any point (barring gravitational singularities). More abstractly, we say that in the presence of gravity space-time is described by a curved 4-dimensional manifold for which the tangent*

> *space to any point is a 4-dimensional Minkowski space. Thus, the structure of Minkowski space is still essential in the description of general relativity.*
>
> *Wikipedia – Minkowski space*
>
> *(01/Aug/2014)*

Apologies to those who do not respect the source of this quotation, but I believe this to be an excellent, clear description.

The concept may seem complicated, but what it suggests is that we can regard space-time as flat (or straight) in the three physical dimensions and with the one variable (curved) time dimension, but only in weak gravity fields. The inference from General Relativity is that strong gravity curves space as well as time. Even so, we can still revert to using Minkowski, "flat" space-time even in strong gravitational fields, but only in very small regions, since any curve looks straight at any point if you consider a short enough length of it.

This seems to me to be a contradiction in terms. On the one hand, General Relativity insists that space-time can be curved in all four dimensions, yet it allows the use of flat space time (flat in three dimensions), in any solution within both weak and strong gravity fields, in very small regions. Crucially, if we apply the Causality Principle to gravitation, then gravitational acceleration must be the result of some proximate, antecedent cause, a cause which can be considered to act at a point, at any and all points. Since these points in question *are* the "very small regions" referred to above, then the *cause and effect* of gravitation can *always* be considered to be within local, Minkowski space!

Again, this begs the question: "Is the distortion of the three physical dimensions really necessary?" Certainly, it seems to me (and apparently to some scientists in relativity) to be unnecessary, at least for Newtonian gravitation (weak fields). The mainstream understanding of Newtonian gravitation (weak fields), aligns with the proposal that the time dilation field is the cause and gravitational acceleration is the effect, all within a flat three-dimensional space and with the time dimension as the only variable or curved dimension.

This chapter will take you on a zigzag journey, jumping between the macro world of planetary interaction and the micro world of the atom. We will go around the moon and back, then down to the very centre of the Earth and back up to the surface. We will learn lessons as we go and apply them in each phase of the journey, until we arrive at our final destination of a clear understanding of the physical cause and effect of Newtonian gravitation.

Cumulative Gravitation

In practice, we know that gravitational effects do combine and are cumulative, if they are acting in the same direction, and that they counteract, if they are acting in opposition directions. The best known example of this is the interactive gravitational effects between the Earth and the Moon. In the space in between the Earth and the Moon, the gravitational attraction towards the Earth decays with increasing distance from the Earth. It is also counteracted by the Moon's increasing gravitation acting in the opposite direction. As we move towards the Moon, a point is reached where the gravitational attractions of the Earth versus the Moon are equal and opposite and the net effect is zero. This "Neutral point" is 43,000 miles from the Moon's surface on a straight line between the centres of the Earth and the Moon. At this point, it is theoretically possible for a body to stay there indefinitely without falling towards either the Earth or the Moon, but you would have to steer around with the Moon in an ellipse, concentric with the Moon's orbit to keep yourself in this neutral position.

The American Apollo spacecrafts slowed down to a mere 2,200 mph, just enough to get them past the neutral point and to then allow the Moon's gravity to take over from the Earth's, on their free fall, onward journeys to Lunar orbit. At this same radius from the Earth, but remote from the Moon, say on the opposite side of the Earth, the same body would fall towards the Earth and not remain there, as there is now no neutral point. At this position there is only the Earth's gravitational field, still decaying towards infinity, and the Earth's gravitational attraction will therefore prevail.

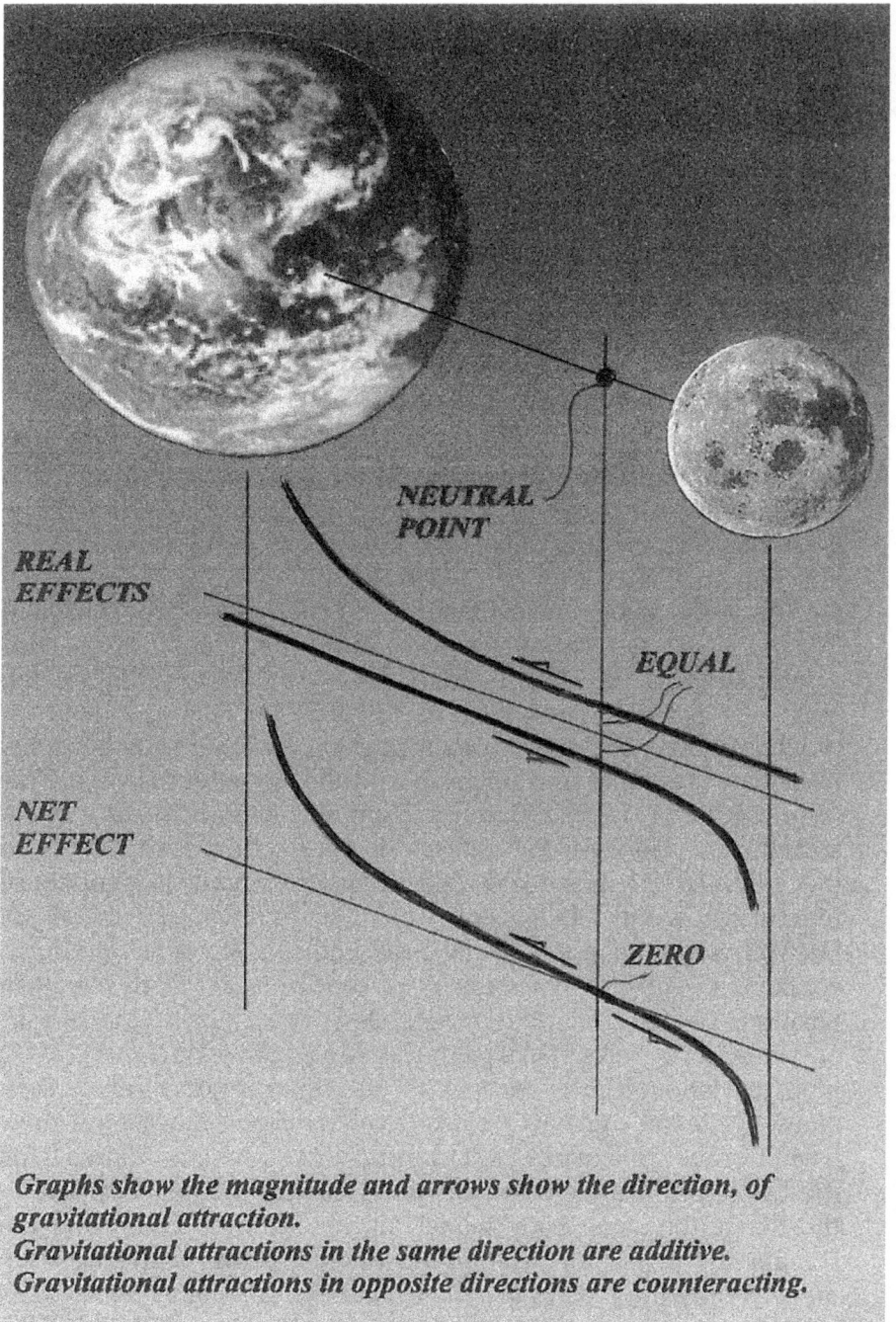

Graphs show the magnitude and arrows show the direction, of gravitational attraction.
Gravitational attractions in the same direction are additive.
Gravitational attractions in opposite directions are counteracting.

Figure 5: Earth and Moon Gravitational Interaction

Figure 5 shows a simplified linear diagram demonstrating how the two fields interact. The reality is of two spherical fields superimposing but we can visualise their spherical nature from the linear graphs in the figure.

In Figure 6, beyond the far-side of the Moon, both the Earth's and the Moon's gravitational attractions become additive since gravitational shielding[47] (and also time dilation shielding), does not exist in nature. A body at this location will experience the full pull of the Earth (as if the Moon were not there), as well as of the Moon and it will accelerate slightly faster towards the Moon's surface than it would do on the Earth-side of the Moon where the fields counteract.

One can visualise how the gravitational spheres of influence, of both the Earth and the Moon, interact with one another, both between (and *beyond*) the Moon, as shown in Figure 6.

We can see that, beyond the moon, the gravitational attractions of both the Earth and the Moon are still present and that the net effect is the summation of these. If we had a row of planets, all in line, then the net gravitational attraction, beyond the last planet, would be the summation of all the gravitational effects, from all the planets, at their respective distances. The planets are "transparent" to the gravitational and time dilatational fields.

---

[47] Gravitational shielding – The hypothetical effect, where the gravitational pull of a body (Earth), is prevented by the presence of another body (Moon), in the "shadow" of the other body. This does not exist in nature.

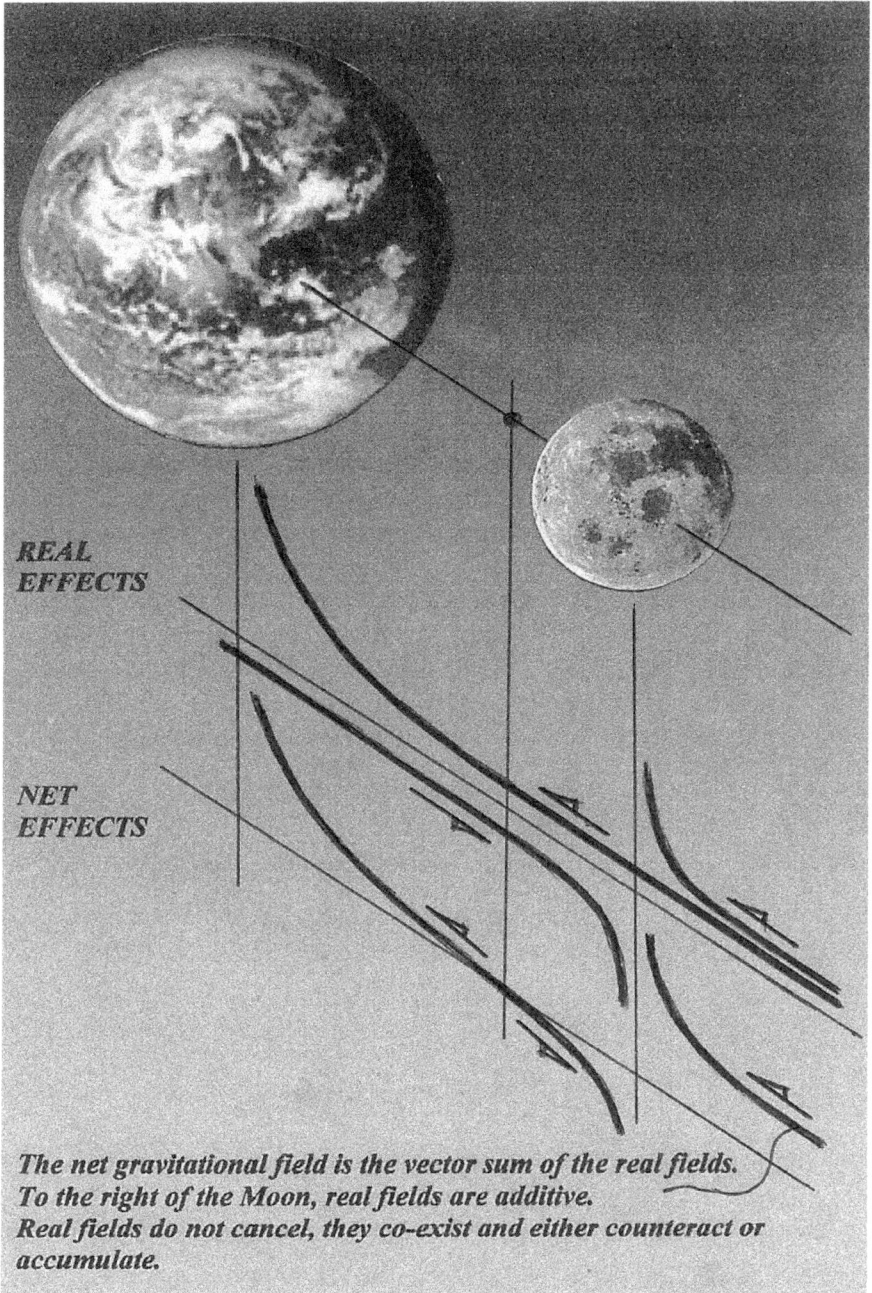

The net gravitational field is the vector sum of the real fields.
To the right of the Moon, real fields are additive.
Real fields do not cancel, they co-exist and either counteract or accumulate.

Figure 6: Earth and Moon gravitational accumulation

The Equivalence of Time Dilation and Gravitation

We must now pursue the idea of a direct link between the time dilation caused by the presence of say, the Earth, and the gravitational effects also seemingly caused by the presence of the Earth. I use the word "seemingly", since we can only *assume* that time dilation is directly caused by the presence of mass. Similarly, we can only *assume* that mass directly causes gravitation and only because both effects are always coincident with the presence of mass. We should be aware that we have become so used to associating gravity with the presence of mass that we have come to believe that there is a *direct* causality between mass and gravitational acceleration, but even today, this is not scientifically conclusive.

We know that both time dilation and gravitational acceleration decay, with distance from a mass, in accordance with the inverse square law, or similar shaped curve of decay.

We know that gravitational acceleration (at any particular distance) is proportional to the mass of the large body.

We know that the time dilation (at any particular distance), induced by a massive body, is also a function of the mass of that body.

We can therefore deduce that the time dilation (at any particular distance) can be expressed as a function of the gravitational acceleration (at that distance) and vice versa.

This deduction *is* scientifically conclusive and we can therefore make the following statement of fact:

*Newtonian gravitational acceleration is a
function of time dilation*

This verifies that both time dilation and gravitational attraction are one and the same thing or perhaps that one directly causes the other.

In support of this deduction, time is the only known entity in a vacuum in a gravitational field (on the macro scale) and if we allow Occam's razor to guide our thinking, we shall avoid creating any new entities.

If we were to deviate from the simplest course and create new entities like the graviton, in order to explain how gravity might work, then we would diverge from the simplest explanation, without ever having disproved it.

There has been a reluctance to stick to the guidance of Occam's razor and we have preferred to invent the spurious graviton, rather than explore the simpler alternative. This is where science has failed from the early twentieth century until now.

If we maintain our discipline and refrain from risky creative thinking, which might generate misleading concepts like the distortion of space, or even the graviton, then we have to believe that the only entities (and effects) in the void are, the passage of time (at whatever rate at any point) and the observed ultimate effect of gravitational acceleration. From this, we can make the following logical deduction:

> *Since acceleration due to gravity is the ultimate effect and, since space cannot be a cause for anything (because space itself is nothing and has no attributes), then, in the absence of any newly created entities, we are forced to conclude that it must be the entity of time which is somehow the cause of the observed gravitation.*

Bearing in mind all the above, our initial assumption that time dilation and gravitation are one and the same thing, or that one causes the other, is now a *conclusion* that time dilation somehow causes gravitational acceleration.

This conclusion rests on the initial assumptions that there are no entities (gravitons) within space-time apart from the passage of time and that space itself is flat. The only thing missing of course is a mathematical proof and this will be forthcoming in the next chapter.

Gravitational Shielding

Perhaps now it is worth deliberating on the gravitational shielding issue.

Gravitational shielding is the hypothetical effect whereby a solid object prevents gravitational interaction within its "shadow", beyond the mass causing the gravitational field. Some have proposed that gravitational shielding is a real effect, but there is no available evidence to support this proposal and scientists are, generally, in agreement that gravitational shielding does *not* occur in nature.

Time pays no respect to mass or massive objects like planets (or even stars) and passes straight through them as if they did not exist.

The effects of time curvature are therefore at liberty to permeate macro space indifferent to any matter contained within it. Only when met with another reality such as another time dilatational field, acting either in the opposite direction or in the same direction, will the effect be diminished or increased, but, even then, only locally where the fields interact. When the spread of time dilation reaches a space where there is (virtually) no other field, it is as if there never was any interaction. The time dilation strength is as it would have been, had the other field never existed.

This is a somewhat simplified argument, since all fields reach to infinity, but the point is nevertheless made. No evidence has been found to support the existence of gravitational shielding in nature and, in any case, it would be in direct contravention of Einstein's Principle of Equivalence.

The proposed theory relies on the fact that it is not possible to shield from gravitation by simply screening with baryonic[48] matter (and that the presence of another mass has no effect on the time dilation field from the original mass generating the field).

---

[48] Baryonic matter – Normal matter. Baryons are a group of subatomic particles made from three quarks. They make up the protons and neutrons within the nuclei of atoms.

The following principle, accepted by current science is:

> *Gravitation/Time dilation act as if there were no other entity in the universe and they pass straight through all baryonic matter. Two or more gravitational or time dilatational fields can co-exist at the same location, unaffected by each other, but their net effect at any point is the vector sum of their fields at that point.*

In engineering terms, if two or more vectors exist in the same location, then the *net effect* will be the simple vector sum of these. The fact that we are adding or subtracting *real* vectors to arrive at the *net* effect does not mean that they have cancelled or diminished each other. Both still exist with their initial, individual values unaltered, irrespective of their combined effect. This concept is important for understanding the next sections.

> *Overlapping time dilatational and gravitational fields do not cancel each other. They coexist, and either counteract or accumulate depending on their relative directions.*

This is an important principle to grasp that, although the gravitational effect of, say, the Earth is reduced by the Moon's effect acting in the opposite direction, this does not prevent the real gravitational fields from combining beyond the Moon as if there had been no previous interference. This principle, when applied at the atomic scale is fundamental for understanding how gravity is created.

Time Dilation on The Atomic Scale

Now let us consider the atom, in particular, the spinning of the electron around the nucleus. We might also consider the other fundamental particles that have rotation (or any sort of physical "spin" or momentum). The same principle will apply here as for our global positioning satellites, namely, the time rate for the rotating particles will be slower than that of the "stationary" parts of the nucleus. In the case of the electron, its time rate will be measurably slower, due to the extreme tangential speed of the electron, around 1/150 of the speed of light. Therefore, we might consider that the time rate for the electron's frame of reference is slow, relative to its stationary surroundings in the macro world.

There may be some debate about the validity of applying relativistic time dilation to the motion of electrons and other subatomic particles, since the electron is envisaged as a very indistinct electrical charge, a wave packet or "smudge", if you like, of electrical energy, with a doughnut-shaped realm of possibilities for its orbit. The precise motion of other fundamental particles is even less clear. We need to remember that relativistic time dilation is a property of the mass energy and the frame of reference that the entity has assumed, the time rate at its speed and not just a property of the entity itself. It is irrelevant what form the entity might take. What is important is its mass energy and its speed. We can argue that the electron is electrical charge = energy = mass, and therefore will dilate time, due to its motion, in accordance with Lorentz.

For the purposes of developing the new theory for gravitation, we need to consider our spinning particles as being virtually static, when viewed from a distance or, at least, vibrating just a little at a fixed point. That way, we can visualise them as "point sources" of time dilation. In other words, we will ignore the very small physical dimensions of the atom and concentrate more on its time dilation field. It is useful to understand that we are all in static proximity to the matter around us. We are in the same frame of reference as all the static parts of the atoms in our surrounding environment. But, we are in a *different* frame of reference to all the spinning particles. Their frame of reference has a slower time rate than ours.

Before we proceed further, we need to understand how time dilation propagates through space and we need to answer the following question. When an electron's rotational motion causes it to have a different frame of reference, with a slower time rate than a static frame of reference, then where, exactly, are the boundaries of the time dilation from the moving frame of reference? After all, you cannot have a clock with one time rate, right next to another with a significantly different time rate. This never occurs in nature, there is always a gradual process of spreading outwards and of a decaying effect as the field occupies more volume. Time curvature is exactly that, it is curved and not stepped (at least not at the macro scale).

It is an accepted scientific fact that any point source which spreads its influence equally in all directions (without a limit to its range), will obey the inverse square law. The inverse square law as an example is:

$$g = GM / r^2$$

where for any mass M, "g" is proportional to the inverse of the square of "r".

This example is Newton's law of gravitation, where g is gravitational acceleration, G is the gravitational constant, M is the mass of the body creating the gravitational field and r is the distance of the measured gravitational effect from the point source. This comes from strictly geometrical considerations as shown in Figure 7. The intensity of the influence at any given radius "r" from the source is the source strength divided by the area of the sphere at radius "r". Being strictly geometric in its origin, the inverse square law applies to diverse phenomena. Sources of gravitational "force", electric fields, light, sound, and radiation all obey the inverse square law, so the time dilation effects induced by the motion of mass or energy will similarly show a decaying field.

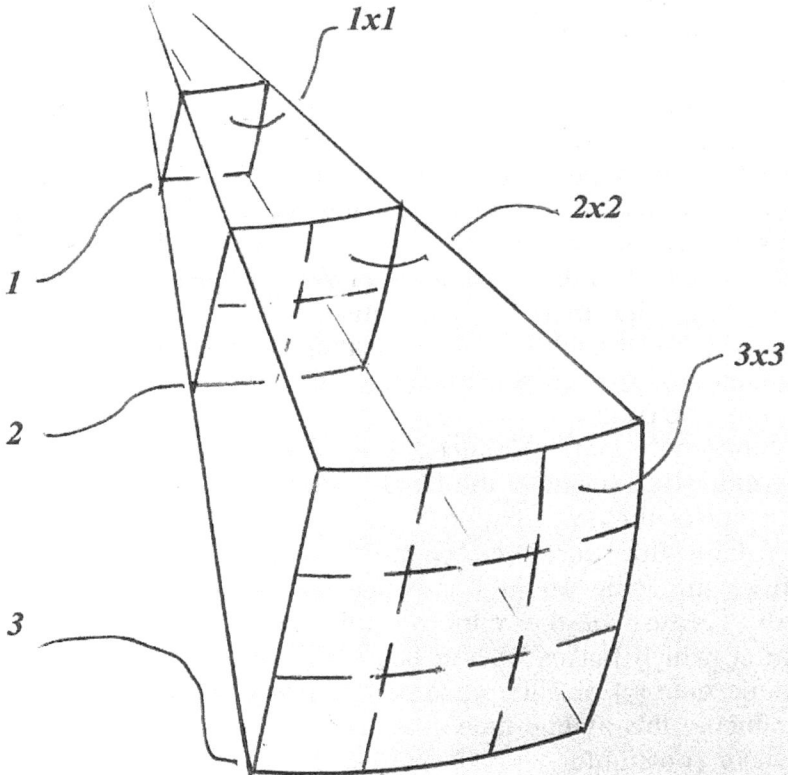

The radiating effect is spread over increasing area as it expands outward from the source.

The energy per unit area at any distance is proportional to $1/r^2$

Figure 7: Inverse Square Law

Cumulative Time Dilation

When we look at the Earth from deep space and wonder why such bodies have a gravitational field, we can only guess. Even General Relativity does not suggest a cause and effect, but simply claims that mass somehow distorts time and space. The new theory proposes that gravitation is purely the effect of time dilation and so our question changes to "Why does the Earth have a time dilational field?"

We know, from deductive reasoning, from mathematical proofs in Special Relativity and from experimental verification, that speed slows time. The time dilation must then decay with increasing distance from the moving source. However, we have no such proofs for *mass* induced time dilation and we just blindly accept that mass dilates time. We have not questioned GR in its declaration that there are two causes for the same fundamental effect, even though double causation is unknown in nature. Ultimately, at the fundamental level, every single effect boils down to only one cause.

Since only motion-induced time dilation is proven, we must therefore look again at the Earth and ask, "Where is the motion that causes the time dilation?"

Asking this question reminds us that there is plenty of movement going on at relativistic speeds within any massive body, because of the spinning subatomic particles within the matter which makes up the body. The internal momentum (or kinetic energy) of each subatomic particle will play its part in producing this atomic time dilation, but electron spin is perhaps a major contributor.

We shall now try to consider the almost unimaginable number of atoms which make up the Earth with their spinning electrons which, due to their motion and the effects of Special Relativity, have dilated the time rate in their immediate, sub-atomic vicinity. The time dilation then decays with increasing distance from each atom in accordance with a version of the inverse square law and we can perhaps now start to visualise the cumulative effect in a large mass.

The Earth's time dilation is produced by the combined, subatomic kinetic energy of the mass of the Earth. Ultimately, it is the accumulation of all the individual time dilation effects, from each spinning particle within each atom, that produces the total time dilation effect at the planet's surface. This total accumulated time dilation effect at the Earth's surface then decays with distance from the Earth, in accordance with some form of the inverse square law. This decaying effect, above the surface, is what we observe in practice, today, as the difference in proper time[49] at various elevations in the Earth's time dilation field. Remember, time dilation is the slowing down of time, so it is always negative in sense. As the effect diminishes or decays with distance, time will speed up as the time dilation reduces.

> *The accumulation of the inertial time dilation fields from the spinning subatomic particles of a massive body, ultimately gives rise to the total time dilation effect at the surface of the body itself.*

This is a new proposal for the fundamental cause of the time dilation field and, therefore, for the phenomenon of Newtonian gravitation. In support of this postulate, when we consider the early universe, before the formation of any stars or planets, we know that individual Hydrogen and Helium atoms attracted each other to form the first stars from which all the elements were created. At that time, there were no clumps of material, no dense bodies to exert gravitation, but only free-floating atoms and so we again deduce from *this* process that:

> *Gravitational attraction on the macro scale is generated within the atom!*

---

[49] Proper time – Local time, as is shown by a clock in the region we are considering. Different in value to the proper time at a different elevation in a gravitational field, or in another frame of reference moving relative to the local frame.

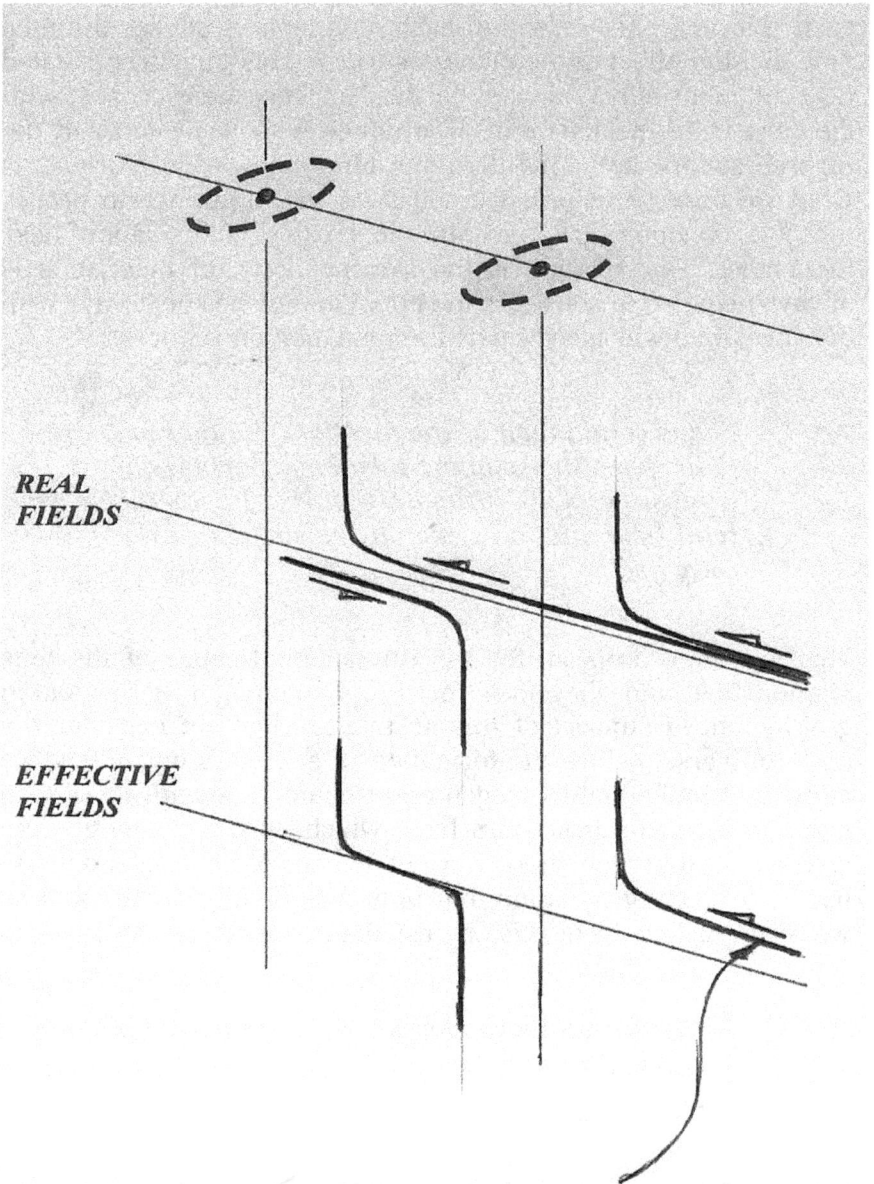

The effective field to the right is amplified by the accumulation of the real fields at that position. This accumulation is unaffected by the previous interaction between the atoms.

Figure 8: Time dilatational interaction between electrons

From this, we see that the time dilation, caused by spinning electrons, not only creates the gravitational effects of massive bodies that we observe today, but that it is the same phenomenon which created these massive bodies, from the free atoms in the primordial void. This concept, of time dilation induced gravitation, gives us a consistent cause and effect, not just for the current epoch, but from the creation of the universe and since the very beginning of matter, as we know it.

Just like the cumulative gravitational/time dilation effects beyond the Earth and the Moon, we can expect a similar accumulation of time dilation beyond any two adjacent electrons. A similar diagram (as that for the Earth/Moon gravitational interaction) is shown in Figure 8, depicting the interaction and accumulation of time dilation between (and beyond) two hydrogen atoms.

If we now consider the fabric of the Earth as a generally homogeneous material, with enormous numbers of electrons in close proximity, we can see, from the following diagram, that the net effect of all the countering and accumulation of the time dilation from all the electrons, ultimately results in the *accumulation* effect *only*. This is because the countering effects are purely local interactions and do not influence the accumulated effects beyond the next electron and so on. Real fields do not cancel, but either counteract or accumulate depending on their relative directions. Remember that gravitational and, therefore, time dilatational shielding does not occur in nature.

Figure 9 shows a single line of electrons, to demonstrate the point. The reality, of course, is spherical rather than linear. Just like the time dilation/gravitational effects between the Earth and the Moon (and beyond), the accumulated time dilation *beyond* any pair of electrons is unaffected by what happens *between* any two electrons, or for that matter, by any other time dilation effect, from any other location. The counteractions are purely local effects, but the accumulation goes on to the extremities of the mass, and then, from the value it has achieved at the surface, it decays in accordance with a type of inverse square law, toward infinity. Above the surface we run out of spinning electrons to generate more time dilation, the field then decays naturally with distance.

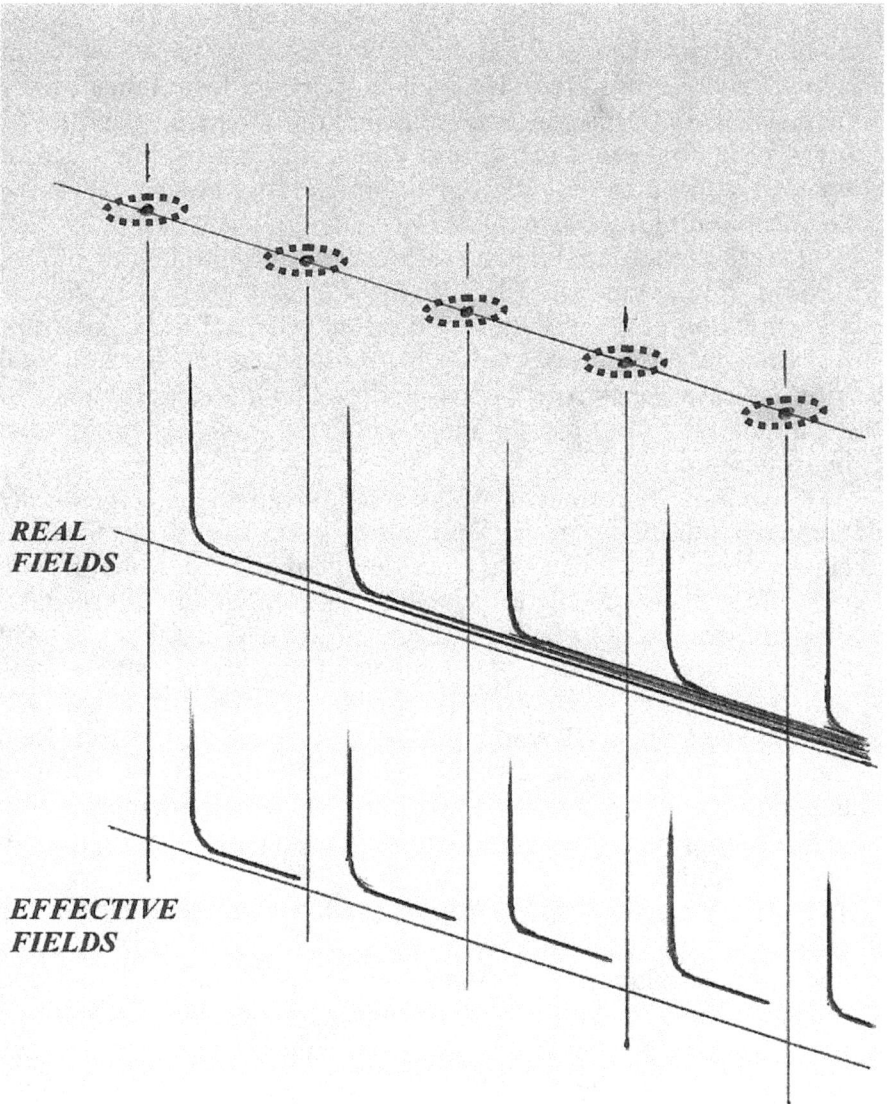

For clarity, the diagram shows a single line of electrons and the time dilation field in one direction only. In reality the field is a spherical matrix with the time dilation field acting in all directions.

Figure 9: Time dilatational accumulation from electrons

Every tiny effect of time dilation, created from the centre of the Earth, starts to decay immediately, as we move outwards, but all the remnants of the real, subatomic fields, at any point, combine with those of other, locally-generated, time dilation, and so on for 6.4 million metres, the Earth's radius. Added to this effect, are the time dilation effects from the other half of the Earth on the far side of the core, albeit a much smaller contribution being further away. Certainly the electron at the centre of the Earth contributes very little indeed towards the planet's cumulative time dilation. But, as you move outwards from the centre, the distance to the surface for the time dilatational decay becomes less and the number of contributory electrons increases dramatically. It is, therefore, the outer layers of the Earth that do more of the "work", creating the time dilation at the surface. Nevertheless, the outer layers require that the inner layers exist, so the time dilation at the surface is a function of the whole mass and not just a specific part of it.

Regarding accumulated effects acting in the opposite direction, remember that the real fields of time dilation do not cancel, but co-exist and counteract. So, for every opposite accumulated effect, right through to the other side of the Earth, there will be no diminishing effect on the accumulated end result, in the direction we are considering, or indeed, in any direction. Considering the huge mass of the Earth producing this effect and considering the minute effects of the decaying time dilation from each atom, it is perhaps not surprising that, in practice, the accumulated time dilation of the Earth has only reduced the time rate at the Earth's surface, to a value of 0.9999999993044123[50] of the time rate in the otherwise empty space at Earth orbit. In other words, it has taken the huge mass of the Earth, with its almost unimaginable number of spinning subatomic particles, to slow time down by a miniscule amount at the surface. Nevertheless, this is a very *significant* miniscule amount, as we shall see later.

---

[50] 0.9999999993044123 – This value has been calculated from a well-known equation, developed from the Schwarzschild metric, giving the time dilation at a distance from and outside of a non-rotating spherical massive object.

The Link Between Time Dilation and Gravity

Now, imagine you are in your space suit floating, with your fellow astronaut, a thousand miles above the Earth (with the moon on the opposite side, with negligible gravitational effects to avoid confusion). You look earthwards and see the familiar blue and white globe we now all recognise as man's home in the universe (see Figure 10). Imagine, if you will, the time dilation around the earth, slowing the time rate in its immediate vicinity at the surface, then beyond, with the time dilation decaying with distance, in accordance with some form of the inverse square law shown by the graph. Imagine the time dilation, as it starts at the Earth's surface, decaying as it passes through you and continues onwards, beyond you both, eventually decaying towards zero, as distance approaches infinity.

You can now see that this spherical time dilation field, passing through you both, is a real phenomenon and that you are actually on a time dilatational slope, or gradient, which increases in magnitude (negative) as you get closer to the Earth. Just behind your fellow astronaut (in the figure), on the Earth side of him, there is a region of space with the time rate slightly dilated and with a clock showing time rate t'_2 where 2 is frame of reference 2. Immediately in front of him is a region of space with the time rate slightly less dilated, due to its slightly increased distance from the Earth, with another clock showing time rate t'_1 where 1 is frame of reference 1.

You are both in a time rate gradient and you experience the "curvature" of time. The frame of reference is different for different elevations within the time dilation field, just like the GPS satellites and the clocks on Earth.

What effect, if any, does this have on you, floating in space?

Before we pursue this in detail, it is important to remember that you, in your space suit, always have some motion, whether positive, negative or zero, towards the Earth.

If you are "stationary", your motion towards the Earth is simply zero and zero is a perfectly real and respectable velocity that just happens to lie between positive and negative values.

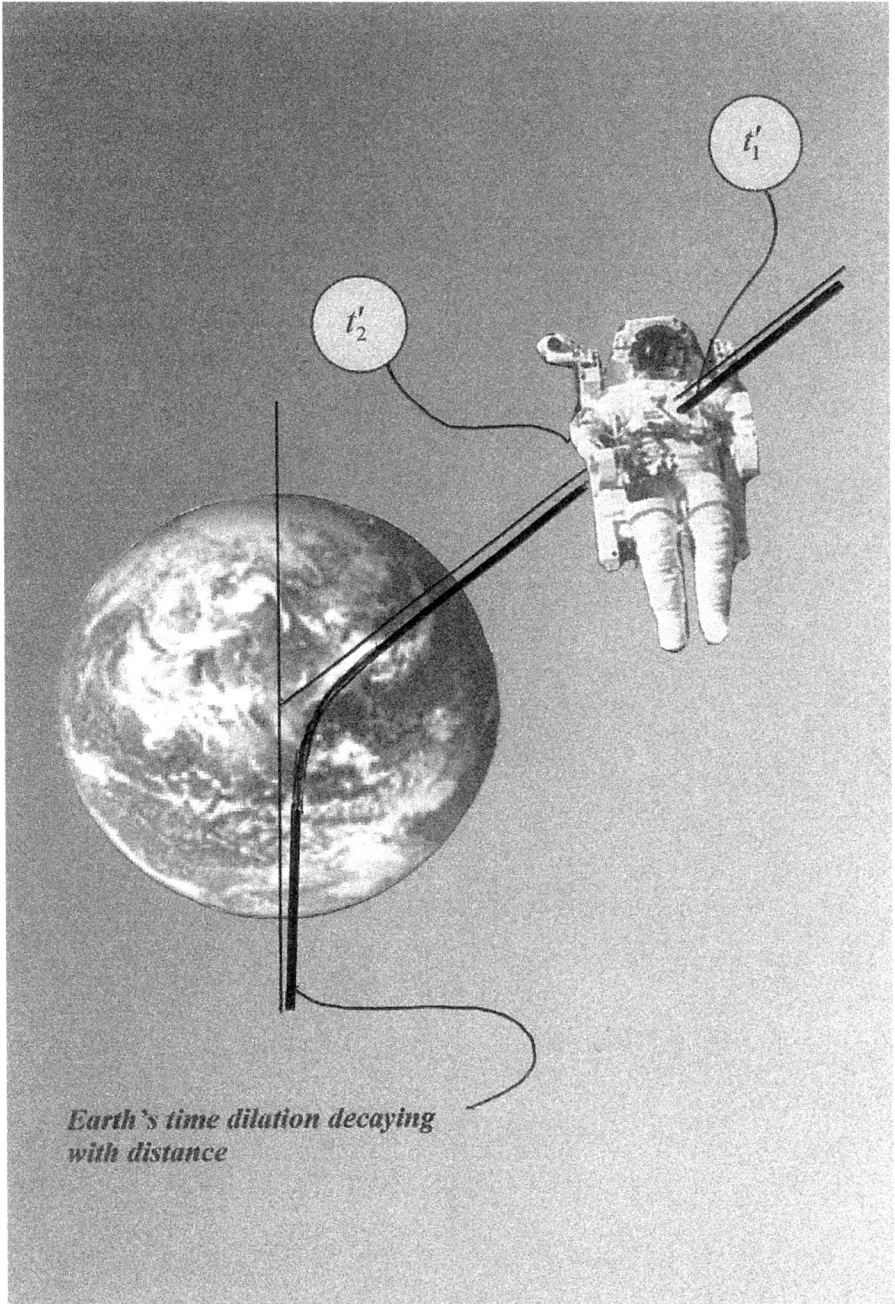

Figure l0: The link between time dilation and gravitation

In other words, we are considering the component of your speed vector, of whatever value, which points towards the Earth.

- We will *set* this component at a constant speed "v" in the direction towards the Earth.

- We will envisage no gravitational field, so that we may see how the time rate variation, alone, affects the speed, in the absence of "gravity".

- It is our intention to prove that the time rate variation will, in fact, provide the gravitational acceleration, without any other cause.

Considering the applied, *constant* velocity "v" in a direction towards the Earth, then we have two simple equations for the points in question (agreeing the convention that motion towards the Earth is negative and away from the Earth is positive):

$$v_1 = -\frac{dr}{dt_1}$$

$$v_2 = -\frac{dr}{dt_2}$$

(Velocity = distance/time, or increment of distance dr / corresponding increment of time dt)

Consider these equations, as viewed by a theoretical observer with a "stationary" clock at a remote distance, unaffected by the time dilatational field. The observer's, hypothetical, remote clock, unaffected by any time dilation, ticks at the rate of 1 second per second of universal time. From this perspective, for each second on the remote, fast clock, a shorter time has passed at clock (1) but an even shorter time still, has passed at clock (2).

Time rate t_1' is faster than time rate t_2'. For each remote time interval dt_f, interval dt_1 will be larger than interval dt_2. Consequently, it now becomes clear, from these simple equations, that v_2 must be greater than v_1 even with our applied *constant* velocity, simply because we have a time rate gradient towards the Earth and so:

$$\frac{dr}{dt_2} \geq \frac{dr}{dt_1}$$

or:

$$v_2 \geq v_1$$

This result, from our simple analysis, shows that acceleration towards the Earth will take place purely due to the time curvature. We can see that the velocity will increase independently of the mass of the accelerating body. You could place an object in the same orbit with any mass, an astronaut, a space station, a satellite, etc., and they would all have the same free fall acceleration. The following scientifically conclusive statement can therefore be made:

> *For any entity in a time dilatational field there will always be acceleration towards the source of time dilation, without any force being applied. The magnitude of the acceleration, at any point, is proportional to the time gradient at that point. This acceleration is independent of the mass of the accelerating body, as it is the result of the time gradient and not because of any applied force. The time gradient at any point (and therefore the acceleration at that point) is a function of the mass of the body creating the time dilatational field and of the distance from it.*

In essence, this is a deductive proof for the cause and effect of Newtonian gravitation and of Einstein's Equivalence Principle. Independently, we have derived the general form of Newton's Law of Gravitation, as well as the Principle of Equivalence, without the use of complex mathematics and purely from this understanding of the temporal cause and effect of Newtonian gravitation. It agrees with accepted science, observation and practical experience, as well as giving a more complete explanation for the cause of gravitational acceleration. It is the only postulate that fully satisfies the causality principle at the fundamental level, at least for weak fields. Our knowledge of gravitation has, therefore, been expanded.

We can see that this acceleration is not forced, therefore no reaction is felt by an individual in this situation. He is in free fall. Both he and his space capsule, for example, accelerate freely at *exactly* the same rate (and he is not pressed to the back of the capsule by any acceleration force). It does not matter what the mass of the accelerating body is, since the acceleration is not due to a force, so Newton's 2nd law[51] does not apply. Nor does gravity have to exert different forces on different masses to accelerate them all at the same rate, if we understand that the dilation of the dt term in the simple velocity equation is the only driver of the induced acceleration. The acceleration is not dependent on the mass in free fall, but only on the time gradient, at its elevation in the time dilatational field. We can now see that even an entity with zero mass, such as a light beam, will also be accelerated in the same way. This is because the time variation, within space-time, is the cause and this "mechanism" does not recognise mass. It is not a mechanical cause and effect, it is a temporal one.

To further help understand why different masses are accelerated at the same rate in a gravitational field, we need to grasp the difference between a change in velocity, due to a change in time rate, and a change in velocity, due to the application of some force. The former requires no energy input, because it is an effect which results from the variable time rates within space-time, whereas the latter requires energy input, in the form of an applied force, to overcome the inertia of the object being accelerated. The Principle of Equivalence still stands, of course, and when we stand on the Earth's surface, our gravitational attraction of $9.807 m/s^2$ is equivalent to us being accelerated "upwards" in space at the same rate.

In more general terms, acceleration due to gravity is the result of following a line of least action, a "World Line" (as Minkowski once put it) or a "Geodesic" (as Relativity describes it). All entities,

---

[51] Newton's 2nd Law – Force = mass x acceleration.

whatever their mass, will have the same acceleration value at the same distance from the Earth, or any other massive object producing a time dilatational field. This acceleration will always be in the direction towards the centre of the massive body. Newton and Einstein say this in the language of mathematics but, until now, science had not fundamentally described *what* gravity is, *how* it is created or *why* it follows the mathematical rules.

The above demonstrates, independently, that gravitational acceleration is dependent only on the strength of the time dilatational field and it bears no relation to the mass of a body in free fall. We can ignore the time dilation produced by the free falling mass, since this is negligible in comparison to the Earth's time dilation, but it does form part of the gravitational attraction, however tiny the effect. This new understanding of gravity has been developed using logical deduction and visualisation techniques, whilst Einstein's equivalence principle was mathematically derived using Newton. So, it is reassuring to see that they both arrive at the same conclusion having come from different directions. Einstein's equivalence principle, in his own words, states:

> *"A little reflection will show that the law of the equality of the inertial and gravitational mass is equivalent to the assertion that the acceleration imparted to a body by a gravitational field is independent of the nature of the body. For Newton's equation of motion in a gravitational field, written out in full it is:*
>
> *(Inertial mass) x (Acceleration) = (Intensity of the gravitational field) x (Gravitational mass)*
>
> *It is only when there is numerical equality between the inertial and gravitational mass that the acceleration is independent of the nature of the body.*
>
> Albert Einstein

Of course, the very notion of "gravitational mass" is absurd. There is no such thing, Einstein invented it, simply to make the rules work. It is not a physical property of matter. It is simply an abstract attribute, created to make the math work and to help describe gravitational behaviour. In the next chapter we will derive an equation which expresses gravitational acceleration purely as a function of time dilation. We will show that this is not just a rule of behaviour, but a universal law which encompasses the cause and effect of gravitation and which also makes predictions in agreement with Newton's "Law" of gravity. Further on, in Chapter 7, we will deal with the deviations from Newtonian behaviour currently described by General Relativity. We will show that the curvature of space is, actually, an effect from the rapidly changing time rate over distance in strong fields. It is the time curvature that is the real, fundamental cause and effect for gravitational interaction in all fields, both weak and strong.

6 The Mathematics of Newtonian gravitation

The Mathematical Challenge

We have demonstrated, with sound, deductive reasoning, that gravitation is the direct result of time dilation. Nevertheless, to establish the theory as irrefutable, we must produce a mathematical relationship that demonstrates the time dilation produced by any massive body will always result in a free fall acceleration of "g" for that body. In other words, we need to prove that the mass induced time dilation, alone, causes the ultimate effect of gravitational acceleration, in accordance with Newton's Law of gravitation. We know that a huge mass is required to produce a very small time dilation effect indeed, which at the Earth's surface gives us 0.9999999993044123 of the time rate in empty space, at Earth orbit. On the face of it, this may not seem enough to produce the fairly significant acceleration value we are looking for, namely, $9.807 m/s^2$ at the Earth's surface. It is clear we are looking for some "amplifier", contained within the fundamental law of gravity, that magnifies the relatively tiny values of time dilation into the more significant gravitational acceleration effects we observe in practice.

The Mathematics

The following equation shows the relationship between the time dilation, caused by a spherical massive body, and the radial distance from it. This equation, derived from the Schwarzschild metric, was produced in the early twentieth century and is well known for calculating the time dilation at a distance from the centre of, and outside, a non-rotating sphere. The equation is also consistent with a flat, or Minkowski, space-time.

Although the equation aligns with the mathematics of General Relativity, it can also be derived from Special Relativity.

$$t = t_f \sqrt{1 - \frac{2GM}{rc^2}}$$

t = time interval between events at radius r (Proper time)
t_f = time interval between the same events on the remote, fast clock at infinity and is taken as unity.
G = the gravitational constant
M = Mass of the Earth
r = radial distance from the Earth's centre
c = speed of light.

The Moon's gravitational field, as well as the Sun's (and that of every other local body), will also slow time down at the Earth's position. However, we are only interested in time dilation *gradient* and the effect the Earth's time dilation field has on objects within the Earth system. The Sun's field is approximately "flat" and sloping towards the Sun at the Earth's position, whilst the Earth's field is spherical and, everywhere, acting towards the centre of the Earth. The Earth's rotation will also have a very small effect, which we shall ignore.

The graph in Figure ll shows how time is slowed down, as you approach a massive body. In other words, time dilation decays with distance from the body. The graph is not the typical shape of an inverse square curve, but is more of a sharper "corner". As the time dilation increases (theoretically) closer to the centre of the Earth, it crosses the $t = 0$ point, where this position is the event horizon of the hypothetical black hole formed by the crushing of the Earth's mass to a diameter of just 9 mm, the size of a peanut. At this event horizon, time stands still and $t = 0$.

Of course, there is no force in the universe that could do this and the peanut-sized black hole is the theoretical situation, invoked simply to show the hypothetical mathematical curve for the Earth's "black hole". The curve *outside* the Earth's surface however, is *real*. This external curve describes the time dilation field caused by either the small black hole *or* the real size Earth, since the gravitational effects are the same at the same distance from the centre of either, so long as the masses are the same and we are outside the radius of the full size Earth.

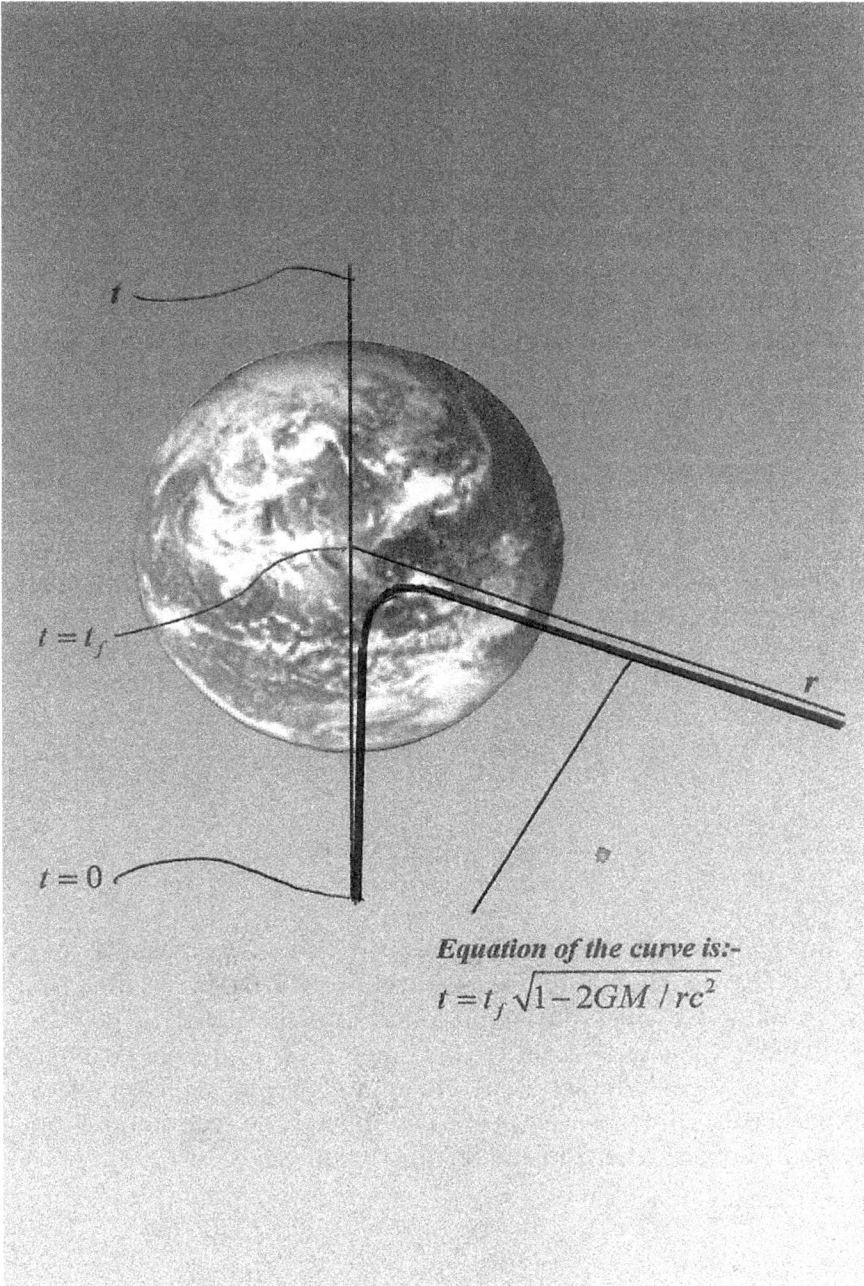

Figure 11: Time Dilation against Distance

Now, Newton's Law of Gravitation states:

$$F = \frac{GmM}{r^2}$$

Where,
 - F is Newton's imaginary "force" to accelerate mass m by "g"
("g" is acceleration due to gravity)
 - G is the gravitational constant
 - m is the mass of the body in free fall
 - M is the mass of the large body producing the gravitational
field
 - r is the radial distance of m from the centre of M

If we replace F, by $F = mg$, (Newton's second law), then we are
simply saying that gravitational acceleration is *equivalent* to a
force on m, in accordance with the law of inertia, $F = ma$
(another version of the Principle of Equivalence, between inertial
and gravitational mass).

Therefore,

$$mg = \frac{GmM}{r^2}$$

and,

$$g = \frac{GM}{r^2}$$

At any radius r, acceleration g is proportional to mass M and
inversely proportional to the square of r. This is verified by
observation and everyday experience (and has been a proven
relationship in science from the seventeenth century onwards). It
is still valid today and accurate enough for predicting trajectories
of spacecraft, satellites and probes or indeed the free fall motion
of any object anywhere in the Solar system.

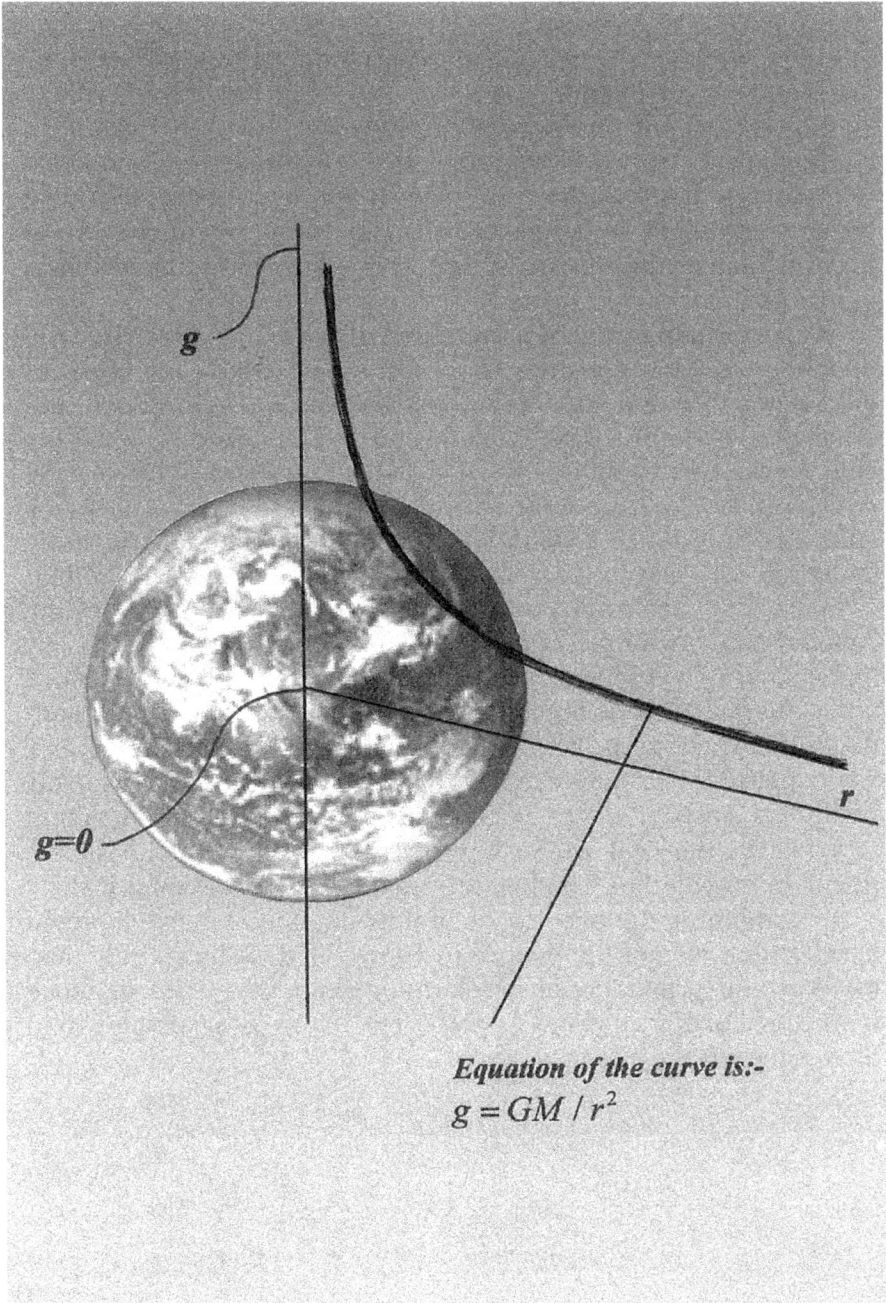

Figure 12: Newtonian gravitation against distance

The new theory regards Newton's gravitational acceleration, as the *effect* of time dilation and Newton's gravitational force, simply as an *equivalent* or *pseudo* force which is only useful in mathematical terms. Nevertheless, Newton's mathematical relationship does define the *behaviour* of objects within a gravitational field, so it does reflect reality in terms of behaviour, if not demonstrating the causality. This is shown by the graph in Figure 12.

Why a pseudo force and not a fundamental law of gravitation? Well, the equation does not contain the cause, only the effect of gravitation. Fundamental relationships contain both cause and effect (e.g. $F = ma$). There is a constant, "G" invoked to make the rule work. This is indicative of a behavioural rule only, a rule which gives the right answers, but which does not describe causality. So, we are looking for a formula that demonstrates the cause on one side of the equation and the effect on the other, without any arbitrary constants involved. This will be the fundamental law of gravitation, *not* Newton.

Now, since we know that the entity of time and the effect of gravitational acceleration are indeed real (and that they both exist in a vacuum at the same time and within the same space), we conclude that they are, in every sense, *simultaneous*. This being the case, it might be interesting to see them in action together in the real world. Figure 13 shows this with a simple linear interpretation. The field effects are of course, spherical.

By combining the two graphs of time dilation and gravitational acceleration we get the picture in Figure 13, this shows how time dilation and gravitational acceleration both vary with distance from the Earth's centre. Critically, the cause is proximate and antecedent to the effect.

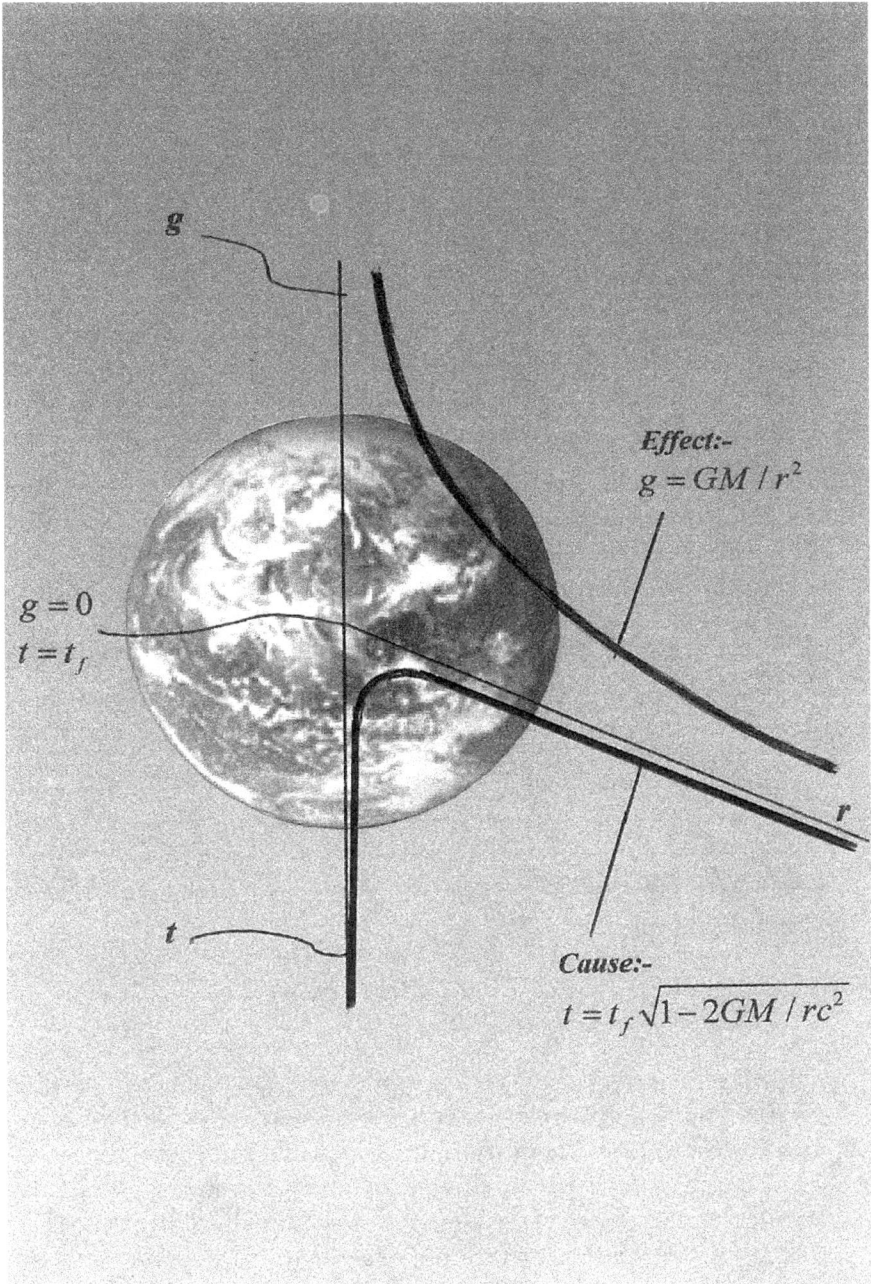

Figure 13: Combined Graph

The two equations are also mathematically simultaneous and are:

$$g = \frac{GM}{r^2} \qquad \text{(1)}$$

$$t = t_f \sqrt{1 - \frac{2GM}{rc^2}} \qquad \text{(2)}$$

Substituting for g into equation 2, we ultimately get:

$$g = \frac{c^2}{2r}\left(1 - t^2 / t_f^2\right) \qquad \text{(3)}$$

This equation tells us that gravitational acceleration, at any particular radius, is purely a function of the time dilation at that radius. We can see that the very small time dilation term is, indeed, multiplied by the very large "amplifier" we were looking for, namely:

$$\frac{c^2}{2r}$$

There are no arbitrary constants.

Note: $-\left(1 - t^2 / t_f^2\right)$ is a factor representative of the amount by which time has been slowed down, the time dilation factor and so we have Equation 3:

$$g = \frac{c^2}{2r} \times Time\ dilation$$

Gravitational acceleration "g" is, therefore, proportional to time dilation, which confirms the conclusion logically derived from Figure 10 in Chapter 5 and the immediate following text.

Equation 3, gives the value of "g" for any mass M, at any particular radius r, purely in terms of the time dilation created by mass "M". Newtonian gravitational acceleration "g" is *not* a result of any force acting at a distance. It is the effect of the variable time rate field, the curvature of time.

The key is that the equation has the cause on one side (time dilation) and the effect on the other (acceleration "g"). Any equation which has these attributes is a fundamental relationship, just like $F = ma$.

Equation 3 is therefore the fundamental equation for the Universal Law of Gravitation (weak fields)

$$g = \frac{c^2}{2r}\left(1 - t^2\right) \qquad (3)$$

Newton's equation is not, and *never was,* the universal law of gravitation. Newton has produced a mathematical relationship, which describes the *behaviour* of gravitational acceleration, but unlike the above equation, Newton does not demonstrate the *cause and effect,* something which Newton himself recognised.

We have used the Newton equation for gravity in this process and some might say that we are therefore bound to prove Newtonian acceleration values, with the above formula, as a result. However, I would point out that this theory is not in disagreement with *any* observation in current science including the behaviour described by Newton. I have simply changed the order of discovery, to define the reality. The starting point for understanding gravitation is not some fictitious force, acting at a distance, nor is it the bending of space itself. It is purely time dilation which directly causes it. Newton's law of gravitational behaviour still stands, of course, from the practical standpoint since his "force" does still "exist" mathematically, but only as an *equivalent* force, a *pseudo* force.

Newton can viably be used here, since we are using the experimentally observed *behaviour* of gravitation which does conform to Newtonian mathematics. We are, therefore, using an experimentally confirmed rule in the development of the new theory. Any attempt to rebut this theory, by disallowing the use of the Newton equation, would be unjustified. To argue that a new theory is invalid, simply because it relies on the observed, proven outcome of the old, but misguided, theory, would be unreasonable. It would also mean that no future progress could *ever* be made in resolving the problems with Newton's force.

We will have lived with the error for too long and become too comfortable with the mistaken idea of Newton's force acting at a distance. We have been stuck at this point for nearly a century of not really knowing what to do about Newton, seemingly lacking either the courage, or the motivation, to clarify this obvious discrepancy with Newton's "force" of gravity.

Relativists do understand that Newton has already been superseded by General Relativity, in that curved space time is regarded as the purveyor of gravitational acceleration and *not* as some force acting at a distance. Having said that, there is still a reluctance to dispense with Newton's force altogether and they also appear to accept the idea of the graviton particle, as well as the curvature of space-time. These are serious conflicts within modern day science. Time dilation is also believed to be a property or an effect of the gravitational field and not its cause, despite widespread acceptance that Newtonian gravitation can be "expressed" in terms of time curvature.

It seems that Equation 3, which states that gravity is purely a result of time dilation, has been available for some considerable time, apparently "hidden", although not very deeply, within our mathematics for over a century. What may have happened, is that this equation has perhaps been regarded as some sort of interim relationship without any real meaning, a by-product of relativity, whereas, I have now shown this equation to be *the* fundamental law of gravitation and Newton's equation is simply another way of describing the resulting behaviour, itself the by-product.

Today, relativists accept that Newtonian gravitation can be regarded as the result of time curvature within an otherwise "flat" space-time. This theory is in full agreement. However, they stop short of stating that time dilation causes gravitation, stating the reverse and that gravitation causes time dilation. This is where we need to change our thinking.

The assumption that gravitation causes time dilation is inherently unsound. Scientists admit that the nature of gravity is not fully understood, so this assumption about what gravity *does*, cannot be relied upon, if we do not know what gravity *is*, in the first place.

Even relativists believe that there is still a force which produces gravitational acceleration. Indeed, the graviton has been invented to suggest a mechanism for the application of this force.

However, any such force is now shown to be fictitious, a pseudo force, so the graviton, as a real particle, becomes unnecessary. Relativity should not require the graviton, since it relies purely on the curvature of space-time to generate gravitational acceleration, yet there appears to be no objection from relativists to this particle's existence. If there has been any objection it has been kept from the public eye. The new theory has set out the space-time model using precise reasoning and deductive logic and the results are in alignment with relativity. Conclusively, the theory deduces that time dilation is the cause of gravitation and provides the mathematical proof. At all stages there is an identified cause for every effect and, in this regard, the theory is watertight.

Causality

The causality principle states:

> *Every effect must have a proximate, antecedent cause*

In other words, if you have an effect, the cause must be local to it and be pre-existing, including any field type cause. The causality principle, applied as a test of the new theory, reveals the following chain of cause and effect.

Stage 1. (Within the Earth)

- Firstly, from Special Relativity, time dilation is caused by the energy of fast spinning electrons and other subatomic particles. This is a proven and observed effect of motion. The effect is the atomic time dilation, the cause is the spinning motion of the particles.

- Next, time dilation decays with distance from the electrons. This is a known, accepted, fact in physics, for any radiating field.

- Then, we have applied the accepted idea that there is no such thing as gravitational shielding in our derivation of the accumulation principle. We have assumed, for the time being, that gravitation is the direct effect of time dilation.

- We have then demonstrated, by conclusive logical deduction, that the time dilation at the surface of a sphere must be the accumulated effects of individual atomic time dilation. The effect is the surface time dilation, the cause being the conclusively deduced internal accumulation.

At each step of this process, the causes are both *proximate* and *antecedent* to the effects.

Stage 2. (Above the Earth's surface)

- From the surface effect of time dilation, we have applied the Schwarzschild equation, which describes the curve of time dilatational decay with distance. This equation has also been used to quantify the surface value. The cause is the surface time dilation and the effect is the decaying spread of this entity in accordance with the Schwarzschild equation. This spread is the spherical time rate field.

- The theory proposes that the time dilation field is the cause of the Newtonian gravitation. The conclusive logical deduction for this is given in Figure 10 with its associated text in Chapter 5.

- We have identified the Schwarzschild and Newton equations as being simultaneous and produced Equation 3.

- Equation 3, shows that gravitational acceleration is, indeed, a function of time dilation only. This relationship validates the preceding proposal that time dilation is the fundamental cause of gravitation.

The basic underlying assumptions are that there are no entities in a vacuum other than just the passing of time (on the macro scale) and that there is no physical curvature of space itself. Again, for each step of this process the causes are both *proximate* and *antecedent* to the effects.

The new theory, therefore, fully complies with the causality principle, as well as being logically, scientifically and mathematically conclusive. It requires no new entities for its workings and so, if Occam's razor *were* a principle of logic then the theory would fully comply with it.

Prior to this theory, there has been no sensible proposal for the mechanism of gravitation which complies with the causality principle. Newton's force, acting at a distance, is prohibited by the causality principle and this has provoked the creation of the imaginary graviton to try and resolve the problem. The graviton is still assumed to exist, today, despite the assertion from General Relativity that gravitational acceleration is the result of the curvature of space, as well as time. This remains an unresolved conflict.

The physical bending of space by the presence of mass is an *assumption,* without explanation or suggested cause and effect and the geodesic, although a valid, abstract, mathematical concept is unreal.

The Cause and Effect diagram Figure 14, shows the complete theory in logic form.

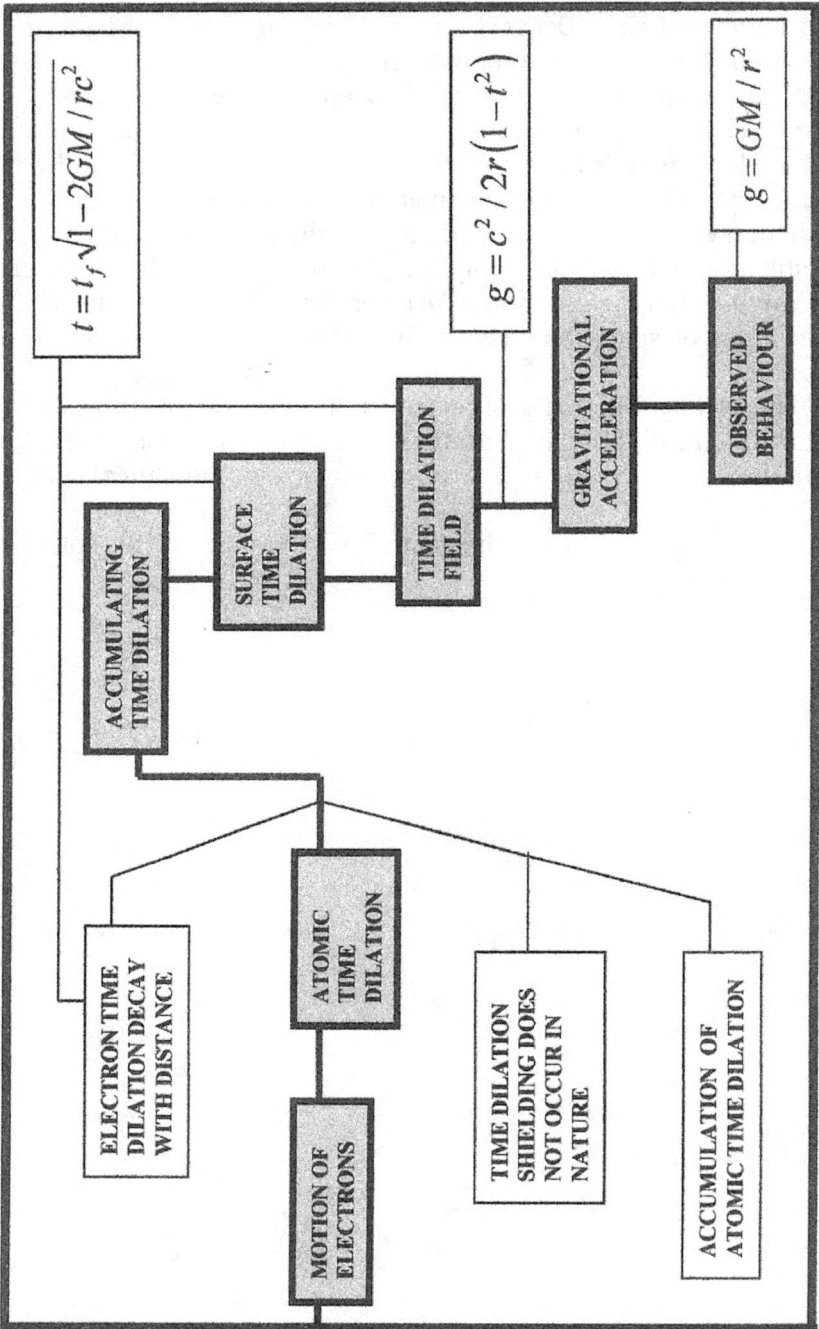

Figure l4: Cause & effect diagram

Presentation

I am not aware of any postulate or theory having been produced, or presented, along these lines, or, indeed, of any individual who views the nature of gravity exactly in this way. Physicists involved with relativity do agree that time dilation is closely related to gravitation. They do accept that Newtonian gravitation can be regarded as the curvature of time within an otherwise flat space-time. Nevertheless, they still talk about time dilation being caused by the gravitational field, or at least by the *red shift* of the field, and not the other way around. They hold on to the graviton as the purveyor of the gravitational force, and also to the curvature of space itself. On all these issues we must now disagree.

- The theory has scientifically concluded that Newtonian gravitation is purely the result of mass induced time dilation.

- The gravitational field is a fictitious creation and is replaced by the time rate field.

- It has eliminated Newton's force replacing it with the direct effects of time dilation.

- It has proposed how mass creates gravity.

- It has provided a watertight cause and effect.

- It has completely unified Newton with relativity, and given a resolution for the discrepancies with Newton's laws.

- It is compliant with at least one exact solution in GR.

Equation 3, is therefore proposed as the Universal Law of Gravitation (weak fields).

Testing the Mathematics

We now need to test the mathematics with a known situation, to see if it gives the right answers. For this purpose, we will calculate the value of "g" at the Earth's surface.

$$g = \frac{c^2}{2r}\left(1 - t^2\right)$$

Note that we have eliminated the term t_f from Equation 3, since it is always equal to unity unless we are making comparisons between local clocks. The remote fast clock always ticks at 1 second per second of universal fast time and t_f therefore equals 1.

A quick test for the sensibility of this equation at its limits reveals:

When $r \rightarrow$ infinity, $g \rightarrow 0$, meaning that acceleration due to gravity approaches zero when approaching infinite distance from mass M.

When r is less than infinity, g will have a value dependent upon the product of $\frac{c^2}{2r}$, (our "amplifier" we were hoping for), and $\left(1 - t^2\right)$, the time dilation factor, which is a very small number indeed even at the Earth's surface.

Due to the opposing scales of these two terms, the accuracy of g, calculated using this equation will be very sensitive indeed to the accuracy of the value we find for "t". It will be necessary to obtain "t" to as many decimal places as possible, and to use a calculator[52] which can deal with sufficient decimal places.

The values we shall use are:

"c" = 2.99792458 x 10^8 m/s

r = 6.3781 x 10^6 m (Radius of the Earth at the equator) and,

r = 6.371 x 10^6 m (Mean radius of the hypothetical Earth sphere)

---

52 Calculator - www.Calculatorforfree.com.

$G = 6.67428(67) \times 10^{-11} \ m^3 / kg.s^2$ (With a standard
uncertainty of 1/10,000)
$M = 5.9742 \times 10^{24} \ kg$

First we must calculate the time rate t at the Earth's surface and
we must use the equation (2):

$$t = t_f \sqrt{1 - \frac{2GM}{rc^2}} \quad \text{-----------------(2)}$$

And so,

$$t = 0.9999999993044123 \ ,$$

This number is a unit-less ratio giving the time rate at the Earth's
surface as a proportion of the time rate in empty space at Earth
orbit.

Now,

$$g = \frac{c^2}{2r}\left(1 - t^2\right) \quad \text{----------------(3)}$$

And so,

$$g = 9.801712835151185 \ m / s^2$$

For Earth radius $r = 6.3781 \times 10^6 \ m$ (Equatorial maximum)

$$g = 9.823332390756828 \ m / s^2$$

For Earth radius $r = 6.371 \times 10^6 \ m$ (Mean radius of the hypothetical
Earth sphere)

The accepted value of "g" at the Earth's sea level surface is
approximately $9.807 \ m / s^2$.

The calculation is an approximation due to the following:

- The Earth is "egg shaped", or oblate, and not a perfect sphere. Gravity is known to be 0.5% greater at the poles due to both the spinning of the Earth and the greater radius at the equator. We can see from the two different results, how sensitive the calculation is to a change in radius, or shape, of the globe. The *effective* gravitational radius appears to be somewhere between 6.371 and 6.3781 million metres.

- Up to 0.01% difference is recorded due to local geology.

- The accuracy of the available input values.

- The limits of operation of the on-line calculator.

- The effects of the rotation of the Earth are excluded.

- The static mass increase of the Earth due to the gravitational field of the Sun will slightly affect the observed value of "g".

Conclusions:

The accuracy of the calculated values, (-0.05%, + 0.16%) from the accepted value, bearing in mind the approximations, clearly validates the new equation.

We should not be surprised that this approach gives us the right answers, since General Relativity shows us that time curvature is one way of looking at Newtonian gravitation (weak fields). The equation used has been developed from Newton and Special Relativity and so we should expect the same answers.

The point being made here, is that time curvature is the direct cause of gravitation, whilst General Relativity stops short of making this claim. I believe that GR avoids making this direct causal relationship, because the deviations from Newtonian behaviour in strong fields also needed an explanation. It was probably already a preconception, during the development of GR, that the curvature of space would somehow be involved in this explanation. The idea of length contraction was already entrenched in scientific thinking by the beginning of the twentieth century and accepted as being somehow real.

At least, the non-reality of it was not considered, so the development of this idea into the three-dimensional curvature of space was not too far removed from the notion of length contraction. It was not a great shock to the thinking of the day. However, I find this a shocking idea and any real, physical curvature of nothing defies belief. Clearly though, it is possible to create a workable theory based on the *abstract* curvature of space, as in General Relativity.

My preconception is that the only available cause for anything within a macro space-time is time itself, or time curvature. So, I must now propose how GR can be interpreted to involve only the real phenomenon of time to explain deviations from Newtonian behaviour in strong fields. The next chapter deals with this.

But before we proceed, I would like to comment on the graviton. The graviton is proposed as the particle that emerges from the field of gravity, but we now see there is no such field. There is only the field of time. I therefore predict that there is no such particle as the graviton and this is demonstrated by the fact it still has not been detected. But, there must be a particle from the field of time and we will discover what this particle is later and name the field from which it emerges. However, we should note that it is the curvature within the field of time that is the cause of gravitational acceleration, not the field itself and so the particle will only reflect the field and not its curvature. It will not be directly associated with gravity. There is no such thing as a graviton and there is no such thing as the gravitational field. There is only the time rate field.

7 Strong gravitational fields

My assertion that gravitation is purely caused by the changing time rates with position, may seem like I am objecting to General Relativity, but this is not the case. I am proposing that, in regions of extreme rates of change of time (i.e. in strong fields), that the effective space curvature is a direct result of this steep time curvature. The mathematics of GR is not under criticism and I do not believe there is any conflict in terms of outcomes. The predictions from GR are always correct, so we simply need to find a new way of understanding GR, in purely temporal terms, whereas, it currently relies partly on abstract, dimensional or geometric terms.

General relativity views time and space as one entity, space-time, but then happily uses the combination of real time curvature plus abstract space curvature to define it. This is a perfectly acceptable viewpoint for predicting behaviour and the theory is quite sound. My viewpoint, though, is different. I understand that time curvature is fundamental, but that abstract space curvature cannot be fundamental, but must be an effect from some other phenomenon. This effective space curvature must be the result of the way in which time is curving in the field. I am forced into this conclusion, because there is only time and space to play with and, if I reject the real, physical curvature of space, but am forced to accept it as a valid mathematical abstraction, then I must conclude this abstract space curvature has to be somehow due to the time curvature. There is nothing else from which to produce the effect.

The effective curvature of any length, or distance, in GR must be solely due to temporal effects, so that when we speak about the curvature of space, what we are really referring to is the *effective* curvature of space, due to changes in relative time rates with position. We must replace the double causation in GR (curvature of space *and* time) with a doubling *effect* from the single cause of time curvature. This doubling effect is, firstly, that the time dilation in weak fields causes Newtonian gravitation in accordance with Equation 3. Secondly, that the rapidly increasing rate of change of time dilation, over distance, in strong fields, causes the relative, temporal distortion of events across distance, producing the deviations from Newtonian predictions.

Figure 15 clearly shows the basic difference between weak and strong gravitational fields. The curve of time rate is, approximately, a very steep, straight line in the strong field, whilst the curve in the weak field is, approximately, a horizontal, straight line. The time rate changes much more rapidly with distance in strong fields than it does in weak fields. In fact, we ignore it in weak fields and only apply it to strong fields by using GR. We can deduce that this difference in time rate change between weak and strong fields must be the cause of the difference in gravitational behaviour within these fields since it is the only difference between the space-times in either field.

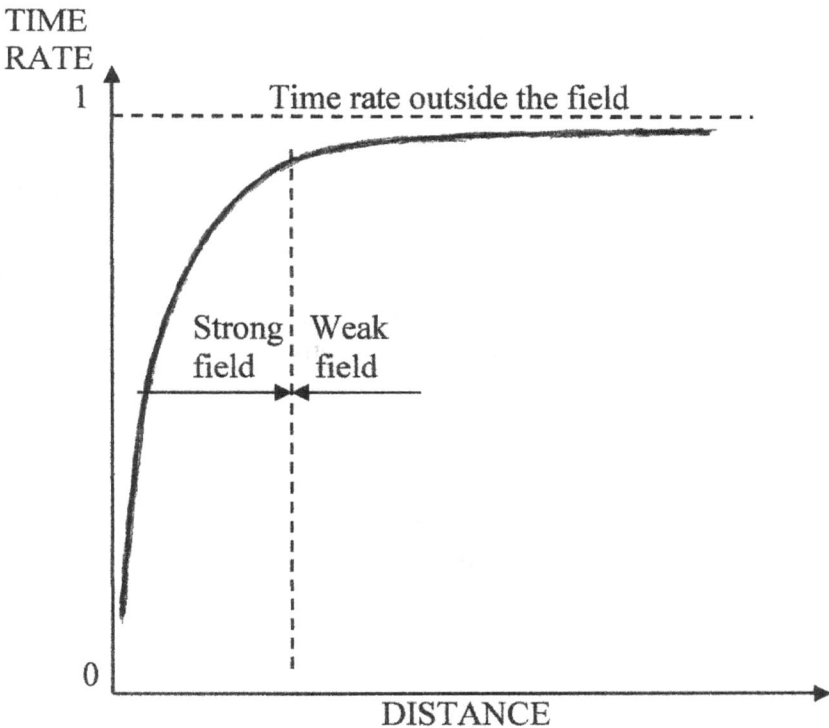

Figure 15: Time rate against distance from a black hole.

*The marked increase in the rate of change of
time rate with distance in strong fields must
be the cause of the effective space curvature,
applicable in strong fields.*

This effect increases the further into the field you go and the
stronger the field gets. The graph shows how time slows down, as
you approach a black hole. We see that, far away, in the weak
field, it slows down very gradually, over distance. It is,
approximately, a straight line in the weak field, where Newtonian
effects dominate and equation (3) below applies, to a close
approximation:

$$g = \frac{c^2}{2r}\left(1-t^2\right)$$

We can see, from this equation, that it is not simply the time
curvature that causes Newtonian gravitational acceleration, but
also the degree of time dilation compared to outside of the field,
i.e. It is the term $\left(1-t^2\right)$ that drives g as well as g being inversely

proportional to radius. The number "l" in the bracket term
represents the time rate outside the field which is taken as unity.
"t" is simply the time rate at any point inside the field shown by
the curve in Figure 15. So, the factor $\left(1-t^2\right)$ is representative of

the time dilation, the amount of time rate lost at any given point
in the field, or the lost temporal energy. Newtonian gravitation is
related to this factor, as well as to the inverse of radius. As you
get closer in to the event horizon, the time rate starts to change
much more rapidly over distance and the curve drops steeply,
quickly falling to zero at the event horizon. This is the "strong
field" where GR uses the curvature of space to explain behaviour
which is now deviating increasingly from Newtonian predictions.

I am asserting that it is this increasing rate of change of time
rate over distance that produces an effect equivalent to space
curvature. The rate of time now changes to a much greater extent
than in the weak field over the same distance travelled and this
effectively makes space, distance, or length look smaller to an
outside observer. It is my intention to prove that space curvature,
which means distances get shorter over distance travelled, and
the rate of increase in time curvature which means that the time

rate gets increasingly slower over distance travelled, are both the same thing. They are equivalent and the deductive proof is very simple.

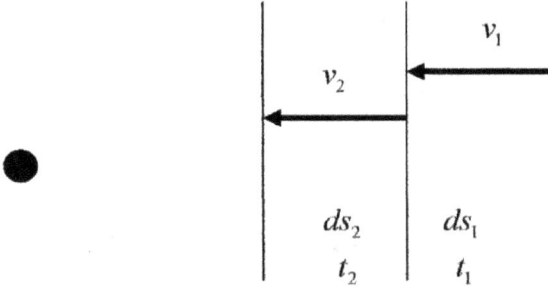

Figure 16: Causality for space curvature

Figure 16 shows two consecutive velocity vectors moving in the direction of, and close to, the black hole on the left. We are considering speeds v_1 and v_2 to be the Newtonian speeds defined by Equation 3 moving through two adjacent small increments of distance ds_1 and ds_2. We are in the strong field so, because of the now significant rate of change of time rate over distance, the time rate, itself, is now significantly different between these two small incremental distances and this can no longer be ignored, as it is in weak fields. This is reflected by the different time rates t_1 and t_2 where t_1 is now significantly faster than t_2. The two equations defining these two vectors are:

$$v_1 = \frac{ds_1}{dt_1}$$

and,

$$v_2 = \frac{ds_2}{dt_2}$$

Transposing these, we get:

$$ds_1 = v_1.dt_1$$

and,

$$ds_2 = v_2.dt_2$$

Now, time increment dt_1 is greater than dt_2 since time is passing at a faster rate in frame (1) compared to frame (2), ($t_1' \geq t_2'$). To define the effective distances ds_1 and ds_2, the Newtonian speeds v_1 and v_2, (which in the limit are the same speed), are multiplied by the different time increments, within their respective frames. Consequently, we can see that $ds_1 \geq ds_2$.

In other words, space has effectively "shrunk" from one frame to the next, because of the significant rate of change in time rate between frames.

The time rate curve in Figure 15, is getting much steeper in the strong field, so this effect gets magnified the closer you are to the black hole. Space effectively "shrinks" even more, as you move forward. Ultimately, at the event horizon, space has shrunk to zero dimensions and, in effect, becomes a spherical hologram around the surface of the sphere at the event horizon. We can now see that space "curves" (gets relatively smaller with position), due to this increasing rate of decreasing time, as you get closer in to the source of time dilation.

Because of this relationship between the ds terms and dt terms, we can see that it does not matter if you use the curvature of space Δds or the curvature of time Δdt, to develop GR, and they must both, ultimately, produce the same results since they both result from the same phenomenon and are directly related. This demonstrates that the abstract curvature of space, (which cannot be a fundamental physical effect), is equivalent to, and *caused by*, the time curvature, which *is* a fundamental effect. Either can be applied mathematically to the Newtonian effects and it will make no difference. The Newtonian acceleration, which is present locally in both weak and strong fields, is amplified by the increasing rate of change of time rate, or the reducing "size" of space, the deeper into the field you go. It makes no difference which we use. I note that this idea keeps cropping up as I write the book, a clue that space emerges from time. If you slow down time, then you shrink space, but if you speed up time, you expand space. In the limit when there is no time, then there is no space, as a result, but we will elaborate on this later. This is not a disagreement with GR, but more, a difference of opinion, as to what to treat as fundamental and what to treat as emergent.

GR takes both the geometry and the time literally and so it takes space as being equally as fundamental as time. I assert that time is *the* fundamental and that space is emergent from time, so I deduce that all effects, within the field, are ultimately caused by changes in time rate.

It is important to understand that Newtonian gravitation is caused by the difference between the local time rate and the time rate outside the field, but the curvature of space is caused by the adjacent differences in local time rate.

Simply put, Newton uses a changing dt/r and GR adds the effect of a more rapidly reducing t'.

I have no doubt that Albert Einstein must have viewed these ideas from all angles, before settling on the best, most practical and most efficient approach to take in developing his theory. He took ten years to do this, after all, and he must have been aware that there would be no difference in results from this temporal approach and his geometric approach. Being mathematically inclined, he chose the geometric route, the simplest route. I do not believe that Einstein missed anything, but his chosen method has subsequently misled others into believing that the curvature of space is a real physical thing. It is of no surprise to me that Einstein took the route he did. If I were in his shoes, trying to develop an accurate, predictive theory, I would certainly have been relieved to avoid the temporal route and to take the simpler, but abstract geometric route. He must have known that they both give the same results after all.

The tools of the trade at the turn of the twentieth century were mainly geometric (and still are in the mainstream physics of today), but our fixation on space curvature must now change. To really understand gravity, we must understand that the curvature of space is merely an effect from the changing time rate and that time is a physical field of energy, the only source of energy we have. Space, or should I say void, by itself, has no energy. It is time that gives space-time its energy.

The next section attempts to show how GR could be made up mathematically from purely temporal terms and without the abstract, geometric space curvature. It quickly becomes obvious that the temporal approach is mathematically more complex, or at least, philosophically more demanding.

It is a much less direct route to the theory, even though it demonstrates a stringent causality while the geometric approach relies on an abstraction.

So, "Why is the temporal approach even worth investigating, if the geometric approach works well enough?" Well, it is important that we never lose sight of causality in science, one real event always causes another real effect. Only by grasping this and doggedly hanging on to it, can we ever hope to truly understand the real, physical world.

The temporal approach clarifies the conflict between space and time, a common discussion in certain articles in popular science publications. As an example, in a recent 2015 article in "New Scientist" magazine, the question was posed by the headline, "Space or Time – One has to go, but which one?" It becomes clear, without too much thought, that all you need to create a space-time, is *time*, just time, *only* time. From both SR and this temporal approach to GR, we see that as the "clock" slows, dimensions effectively get smaller and when time stops altogether, dimensions become zero. They become meaningless and there is no longer any space. Without time there *is* no space. The obvious conclusion is that time creates space and that space is merely emergent from the passing of time. This will be expanded upon later.

Up to this point, in this Chapter, the content is new in this 2nd edition and does not appear in the first edition of the book. A short paper with the following contents from the first edition was submitted to "Physical Review Letters, (Phys.Rev.Lett) on 30th September 2014, but was rejected. The reason given was that the mathematics in the following section did not prove that this approach will give the same answers as GR. It seems I was expected to rework the whole of GR using the temporal approach, something Einstein himself avoided. However, in view of the arguments I have now presented, up to this point, this complete mathematical rework becomes unnecessary to prove the equality of the two approaches. I have clarified the deductive reasoning behind this temporal approach and applied it to the initial equation in the mathematics of GR. It is now obvious that the following mathematics, based on this same logic, is essentially the same as that used in GR and it is *bound* to give the same results.

General Relativity

$$R_{\mu\eta} - \frac{1}{2} g_{\mu\eta} R + g_{\mu\eta} \Lambda = \frac{8\pi G}{c^4} T_{\mu\eta}$$

The above are the Einstein field equations. There are sixteen combinations of $\mu\eta$ but six of them give the same result and so they boil down to ten field equations. For those who are familiar with these, then no explanation is needed. For the remaining majority, it is sufficient to understand what these equations mean generally, in layman's terms:

> *On the right-hand side, the total energy of a gravitational field (including all forms of energy like momentum, stress or pressure), is equal/equivalent to the left-hand side, the distortion effects on space and time caused by the field.*

The equations are an energy balance and give something like a cause and effect for gravitation but, although the mathematics is rigorous, the causality is not. The geometric aspects of the distortion of space are *not* a fundamental form of energy. They are merely equivalent to energy, so there is a fundamental discrepancy within the equations.

   Time however, *is* a form of energy. In fact, it is the *only* form of energy, the fundamental form of energy, as we shall see later. So, any development of the equations, to align them with reality, rather than just with an abstract notion of physical curvature, needs to replace the geometric aspects of physical curvature with the equivalent temporal distortion effects on vectors. Only then will these equations give a true cause and effect. Only then will they be a fundamental law of gravitation and have true meaning in the real sense. This might then allow Mathematicians to complete the temporal theory of gravitation based on reality, on the field of time.

The very idea, once realised, may even allow us to avoid the "disconnect" with quantum mechanics and to ultimately enable us to produce a "Theory of Everything".

It took Albert Einstein ten years to develop General Relativity and he had access to and assistance from, certain professional mathematicians along the way. Clearly then, I am not in a position to rework the whole of GR but let us see if we can find some direction in which it might head, based purely on the phenomenon of time.

Within the initial workings of GR and when describing the physical curvature of space, we come across this equation:

$$d\Phi = \sum_n \frac{\partial \theta}{\partial x^n} dx^n,$$

Where n has values 0, 1, 2, and 3 for the four dimensions of "space-time".

Written out in full:

$$d\Phi = \sum \frac{\partial \theta}{\partial x^0} dx^0 + \frac{\partial \theta}{\partial x^1} dx^1 + \frac{\partial \theta}{\partial x^2} dx^2 + \frac{\partial \theta}{\partial x^3} dx^3$$

where 0 represents time and 1, 2 and 3 represent the three dimensions of space. This equation means that the overall, net *change* in position $d\Phi$, of an entity moving through a curved environment, equals the summation of all the changes in position in all "n" dimensions due to the curvatures in all "n" directions. One does not need to be a mathematician to understand this. It is somewhat obvious. From the equation, the gradient in the "x" direction $\frac{\partial \theta}{\partial x}$, times the distance moved in the "x" direction dx, equals the change in position $d\Phi$ due to the curvature (gradient) in the "x" direction. This is like the change in height from moving along say a 1 in 10 gradient. If you move 5 metres then you drop 0.5 metres in the process. In GR, this method is applied to all dimensions including time, since time is indeed curved in a gravitational field.

Time curvature is real and we can directly observe this curvature, as the changes in time rate, at varying elevations in the field above the Earth's surface (e.g. for the Global Positioning System).

We can understand that, since time does vary for different positions within the field (is curved), then, realistically, we are able to treat time in this way (i.e. when $n=0$). But, and this is a very big "But", we *cannot* realistically treat *space* as curved if we are to stay within the realms of reality. Mathematically of course, we *can* do this, treating physical curvature as abstract. If we are rigorous in our mathematical workings, we will still get to the right answers despite the workings being in the "abstract" world. In reality though, the changes in position, due to this imaginary physical "curvature", are due to the rapidly changing curvature of time, not the curvature of space. Therefore, we ought not to use the same product between physical gradient and distance moved, to establish the effects of "curvature", when n = 1, 2 and 3. The philosophy of General Relativity is abstract, or unreal, in this regard and we should envision reality a little more clearly before we put the math to it.

Again, as with Special Relativity, it is more realistic to focus on the effects of time curvature on time related events, like velocity, rather than on length. Velocity is the ratio length per unit time, so if the time rate varies, then so does the velocity, but, as a consequence, so does the effective length, or distance covered, per unit time (unit time outside the field). Only in this way are lengths *effectively* changed within space-time, but these changes (curvature) of lengths are not causal, they are the *effect* of time curvature.

As a reminder, no one, not even Albert Einstein, nor indeed anyone since, has ever identified a cause for the physical curvature of space. We can only infer it from this *effective* changing of dimensions caused by the time curvature. Time curvature, on the other hand, *is* directly observed on a daily basis, so it is the only available real cause for all effects.

So, what needs to be done with General Relativity? Well, we must modify the geometric terms associated with the curvature of physical space and re-present them as equivalent, temporal terms, reflecting the changes to the energy of time. In this way, the field equations become a true energy balance, a statement that all the positive, field energy of momentum, stress, etc., is equal to the negative time dilation energy of the field and *that* is what General Relativity *is*, in reality, (void of any abstractions).

> *The total positive energy of a gravitational field (including all forms of energy like momentum, stress or pressure), is equal to the negative energy of the time dilation within the field.*

This is a true energy balance and not some impossible attempt to balance energy with geometry (accepting that the changes in geometry are equivalent to the effects of the time dilation).

So, we will start, using the same initial equation from GR since it does give the right answers.

$$d\Phi = \sum_n \frac{\partial\theta}{\partial x^n}dx^n$$

Or:

$$d\Phi = \sum \frac{\partial\theta}{\partial x^0}dx^0 + \frac{\partial\theta}{\partial x^1}dx^1 + \frac{\partial\theta}{\partial x^2}dx^2 + \frac{\partial\theta}{\partial x^3}dx^3$$

This is the fundamental equation for GR from which the field equations are generated. If we can rewrite this equation in purely temporal terms, replacing the terms representing changes in dimensions with their equivalent terms representing the corresponding changes in time rate, then it will remain the same equation.

Both versions of this equation will, therefore, lead us to the same field equations, but one will be Einstein's geometric equation and the other will be the equivalent temporal version giving the same results. We do not have to complete the mathematics to prove that they will give the same answers since both the temporal version and the geometric version are produced from the same initial equation.

The geometric terms with integers 1, 2 and 3, must now all be transformed into equivalent, temporal terms. The first term (time), should stay the same of course. We are considering changes in position due to the *effective* gradient in each direction caused only by the *time curvature* in each direction. This temporal gradient will be different for each direction, since the field is spherical in nature. We could be travelling radially outwards in the field like a rocket taking off, or across the field like a satellite in orbit, or at some compound angle through the field, which is the general case. In each case, the change in time rate with distance is a different value depending on direction and so the gradient varies with direction.

So, for n = 1, 2 and 3, the term $\dfrac{\partial \theta}{\partial x}$ needs to become the *effective* change in position due to the change in time rate over that same distance covered dx, rather than a change in position due to physical curvature or gradient.

If we are travelling with velocity $\dfrac{\partial x}{\partial t}$, or $\dfrac{\partial x}{\partial x^0}$, over a distance of dx, then the distance covered with zero time curvature is simply the velocity times the time taken to traverse the distance,

$$dx = \frac{\partial x}{\partial t} dt .$$

But, since there *is* some time curvature, (time varies over the distance covered dx), then to get the change in position, purely due to this time curvature, we need to multiply dx by the effective time gradient, or curvature, in the direction of travel. Just like with the 1 in 10 gradient. The formula which describes the time curvature in the vertical (radial) direction is:

$$t_r' = t_u' \sqrt{1 - \frac{2GM}{rc^2}}$$

Where t_r' is the time rate at radius r and t_u' is the universal time rate, or the time rate outside the field which we can take as unity.

So, effectively,

$$t_r' = \sqrt{1 - \frac{2GM}{rc^2}}$$

and as r increases, so does the time rate. This gives the change in time rate over distance travelled vertically, or radially, and so the curvature, or gradient, of time rate for any change in height or radius equals:

$$\frac{\partial t}{\partial r} = \frac{\partial}{\partial r}\sqrt{1 - \frac{2GM}{rc^2}}$$

Now, the time increment over the increment of distance dx in any direction x along a path at an arbitrary angle alpha α, to the vertical is:

$$\frac{\partial t}{\partial x} = \frac{\partial t}{\partial r}dxCos\alpha$$

Substituting for $\dfrac{\partial t}{\partial r}$ then:

$$\frac{\partial t}{\partial x} = \frac{\partial}{\partial r}\sqrt{1 - \frac{2GM}{rc^2}}dxCos\alpha$$

Also, from our previous equation for distance covered for zero time curvature,

$$dx = \frac{\partial x}{\partial t}dt$$

This now becomes the following *effective* distance covered, adjusted for the effects of the change in time over this same distance dx:

$$dx = \frac{\partial x}{\partial t}\left[dt + \frac{\partial t}{\partial x}\right]$$

Substituting for $\dfrac{\partial t}{\partial x}$ above, we get:

$$dx = \frac{\partial x}{\partial t}\left[dt + \frac{\partial}{\partial r}\sqrt{1 - \frac{2GM}{rc^2}}dxCos\alpha \right]$$

Now, this is where I apply this new temporal interpretation of dx to the initial equation from General Relativity, describing the change in position due to physical curvature. This proves the equivalence of the two approaches. Einstein's equation:

$$d\Phi = \sum_n \frac{\partial \theta}{\partial x^n}dx^n$$

now becomes the following, (substituting for dx above):
(note $\partial x^0 = dt$)

$$d\Phi = \sum_n \frac{\partial \theta}{\partial x^n}\frac{\partial x^n}{\partial x^0}\left[dx^0 + \frac{\partial}{\partial r}\sqrt{1 - \frac{2GM}{rc^2}}dx^n Cos\alpha^n \right]$$

and so,

$$d\Phi = \sum_n \frac{\partial \theta}{\partial x^0}\left[dx^0 + \frac{\partial}{\partial r}\sqrt{1 - \frac{2GM}{rc^2}}dx^n Cos\alpha^n \right]$$

and finally,

$$d\Phi = \frac{\partial \theta}{\partial x^0}dx^0 + \sum_n \frac{\partial \theta}{\partial x^0}\frac{\partial}{\partial r}\sqrt{1 - \frac{2GM}{rc^2}}dx^n Cos\alpha^n$$

where x^0 is the dimension of time, and n has values 1, 2 and 3 for the three dimensions of physical space. The angle α is the angle from the vertical.

This equation is a balance of increments. The left hand side is the combined incremental movement through time and space in the curved environment. On the right hand side, the first term is the rate of change of time multiplied by the increment of time taken to achieve the combined increment. As predicted, this is the same, first term in the GR equation. The rest, the summation term, is simply the temporal equivalent of the next three terms from the GR equation, but expressed as changes in time. In other words, we have expressed the complete GR equation which uses lengths, purely in terms of time. But, it is the *same* equation!

I think this goes far enough to show how General Relativity can be re-structured on the basis of real or fundamental temporal curvature rather than on the equivalent, but abstract, physical curvature.

Newton's rule of behaviour is:

$$g = \frac{GM}{r^2}$$

But the temporal version of this equation is:

$$g = \frac{c^2}{2r}\left(1 - t^2\right)$$

and these both give the same answers when calculating "g".

Einstein's basic rule of behaviour which starts off the theory in GR is:

$$d\Phi = \sum \frac{\partial \theta}{\partial x^0} dx^0 + \frac{\partial \theta}{\partial x^1} dx^1 + \frac{\partial \theta}{\partial x^2} dx^2 + \frac{\partial \theta}{\partial x^3} dx^3$$

But the temporal version of this equation is:

$$d\Phi = \frac{\partial \theta}{\partial x^0} dx^0 + \sum_n \frac{\partial \theta}{\partial x^0} \frac{\partial}{\partial r} \sqrt{1 - \frac{2GM}{rc^2}} dx^n Cos\alpha^n$$

These two equations are in fact, the same equation.

The last three terms in the Einstein equation are each the product of the way overall position changes with distance moved in each Cartesian coordinate, i.e. $\dfrac{\partial \theta}{\partial x}$, times the distance moved dx in each coordinate. These terms are replaced in the temporal version by the product of three elements. The first is the way that overall position changes with time, $\dfrac{\partial \theta}{\partial x^0}$. The second is how time varies with respect to height or radius in the field. The third element depends on the orientation of the chosen Cartesian axes. In both horizontal cases, the angle α equals 90 degrees and so the cosine value is zero, which leaves only the vertical direction to consider. The more general case must include all directions of movement, but if we take the Cartesian axes as being two horizontal and one vertical then the equation becomes:

$$d\Phi = \frac{\partial \theta}{\partial x^0}\,dx^0 + \frac{\partial \theta}{\partial x^0}\frac{\partial}{\partial r}\sqrt{1 - \frac{2GM}{rc^2}}\,dx^r$$

Which brings us back to our previous equation,

$$d\Phi = \frac{\partial \theta}{\partial x^0}\left[dx^0 + \frac{\partial}{\partial r}\sqrt{1 - \frac{2GM}{rc^2}}\,dr \right]$$

where dr is the incremental increase in elevation in the field as a result of movement in any direction θ.

So, this equation means that the combined, total movement in all directions (in time increment dx^0), equals the velocity in any direction θ, multiplied by the bracket term which is the time increment dx^0, plus a correction for the rate at which the time rate changes with elevation in the field. Movement θ is in any direction, at any compound angle to the vertical but this has to be corrected for the varying time rate with increasing or decreasing of dr, the change in elevation in the field.

Note that the terms on the right-hand side of the equation are either velocity, or time. There are no lengths, or distances, except to define how the time rate changes with elevation r.

With no gravitational field, then the total combined movement would be simply the velocity in any direction times the time taken for the movement, just like in normal movement in zero gravity.

If we were to completely rework GR using this equation, then this approach must give the same solutions as those from the traditional, geometrically based GR. As demonstrated within the reasoning in the first part of this chapter, my proposed temporal interpretation is the direct equivalent of the geometric interpretation of space curvature.

It actually *explains how* space becomes effectively curved whereas GR does not. This temporal version provides a real, physical causality for the effective curvature of space whilst GR cannot be explained logically, but only by the abstract mathematics. The mainstream attitude is, "It must be right, because the math says so". I am averse to allowing mathematics telling me what to believe and a physicist must always seek to identify causality. The geometry tells us what is inevitable, but it does not say how. The temporal method also tells us what is inevitable, but it also tells us how.

This temporal equation is the *same* equation as Einstein's initial four-dimensional equation in GR. It has simply been presented, again, in temporal terms instead of geometric terms, but it has been derived from Einstein's equation. It is the same equation and so it must inevitably produce the same results. It is not necessary to rework the rest of GR, in order to prove that GR can be interpreted both geometrically and temporally in this way. We can say then:

> *The effects of real time curvature on time related events like velocity are identical to those obtained using the abstract curvature of space in General Relativity*

Finally, the equation gives a value for distance moved, $d\Phi$, so acceleration will be the second derivative of this with respect to time, $\dfrac{d^2\Phi}{dt^2}$, and so,

$$-g = \frac{d^2}{dt^2}\left\{\frac{d\theta}{dt}\left[dt + \frac{\partial}{\partial r}\sqrt{1 - \frac{2GM}{rc^2}}dr\right]\right\}$$

This equation allows us to calculate the gravitational acceleration "*g*" at any elevation in the field and takes full account of the effective curvature of space, but due purely to the time gradient. This temporal approach resolves the problem of the distortion of nothing and involves the only real entity in the void, the passage of time. Even the strongest advocates of GR will have to admit this view of the effects of temporal distortion must be given at least the same validity as the curvature of space. It is then simply a matter of deciding which view is more realistic and the answer to this is, frankly, obvious.

You may ask, "If GR gives correct answers (and the effects of time curvature are the same as those for space curvature), then what point are you making? Why do we need to change?" Well, we do not necessarily need to change the math of GR at all and we can go on using it, just like we can go on using Newton's equation for weak field solutions. However, we *do* need to change our *understanding* of the theory, since it is used in the development of many other theories today, such as string theory and loop quantum gravity. This new understanding will impact upon the approaches used in these and other areas.

In weak fields, Newton is still the easiest path to the right answers even though the temporal equation gives a more complete causality. Similarly, in GR, Einstein's equations are the easiest way to get to the right answers, despite the temporal approach giving a rigorous causality. The point is, we should not believe that space can be curved as a property of space. It is just that we can mathematically treat space in this way to get to the right answers, but we must not believe our own "propaganda". If we do so, then we get diverted from a realistic understanding of time and space.

We might believe, for instance, that space itself is quantised or granular. We might think that it can expand (by some unknown source of energy). We might even think we can apply GR right down to the Planck scale, but the curvature of time lies on a much larger/slower scale than at the quantum level, as we shall see very soon.

If we appreciate the above analysis and that time is the only "operator" within space-time, then we can understand our world from a more realistic perspective and not just from an abstract, geometric one. This will enable us to better make progress within all the fields of physics, based on a strict causality, with a much greater degree of confidence, and with mathematics confirming rather than directing our thinking. This is the correct place of mathematics in physics.

When analysing the nature of space-time at the Planck scale, instead of considering how space itself might be broken down into fundamental units, we must now accept that the effective changes in space or dimensions are better understood as the effects from the rapidly changing time rate. Thus, space-time cannot be granular in the physical sense, even though it might seem that way. To take this to its ultimate conclusion, we find that the fundamental for space-time is not the Planck length, but it is the Planck time! The direction in which the development of a theory of quantum gravity is currently heading must now change as a result and Chapter 10 takes this idea to its final conclusion.

We might now appreciate that the idea of space itself expanding is only the effective result from the increasing time rate which gives us the same, observed *red shift* of distant galaxies, as well as an intuitive understanding of dark energy.

With time as the only variable in space-time, we can now more confidently analyse space-time at the smallest scales, not at the scale of the Planck length, but at the scale of the Planck time. This is the approach used in Chapter 10 with the most remarkable results.

8 Some implications

Newton's Gravitational "Force"

Now that we more clearly understand the true nature of gravitational interaction, we could, if we wished, finally dispense with Newton's gravitational "force".

However, it will be necessary, from a practical point of view, to keep the tried and tested Newtonian equations. We should understand, though, that the gravitational force is only an "equivalent" force, a pseudo force.

It would be unnecessarily problematic to dispense with Newton's proven mathematical rule of gravitational *behaviour,* in favour of the real mathematics of gravitational interaction, since it is more practical to still use the formula: $F_e = mg = mGM / r^2$.

In Equation 3, the time dilation values have extremely sensitive arithmetic and it is very difficult to obtain accurate results. So, although Newton is a false notion, regarding the gravitational "force", it is, nevertheless, still the simplest and most direct route to the right answers for weak fields.

The relative distortion effects from Chapter 7 and from General Relativity are another matter. Certainly, for space travel within the Solar system, Newtonian calculations have proved sufficiently accurate up until now.

Newton's First Law

Newton's 1st Law perhaps needs clarification, to take account of the new concept, since gravitational attraction is now seen to be the result of the time dilation field, so there is now no gravitational force.

I recall and propose the following developments:

> *A body will continue in a state of rest, unless acted upon by an external force.*

I remember my mechanics lecturer, at Highbury Technical College in 1967, getting very excited about the deficiency in this incomplete definition and he gave us the following, more detailed definition which is widely accepted today:

> *A body will continue in a state of rest **or of uniform motion in a straight line**, unless acted upon by an external force.*

However, in view of the proposed new concept of gravitation being something other than a force, Newton's lst Law might now read:

> *A body will continue in a state of rest, or of uniform motion in a straight line, **or with a path dictated by its position and velocity vector in a time dilation field**, unless acted upon by an external force.*

Or more succinctly,

> *A body will continue along its own world line (geodesic), unless acted upon by an external force.*

This is already in place from GR.

Newton's 2nd and 3rd laws remain unaffected.

Energy Conservation

The law of conservation of energy states:

> *The total energy of a closed system must remain constant*

The new theory explains how gravity works, without applying a force and, therefore, without requiring the huge amounts of energy to sustain it from all directions, indefinitely. The origins of gravitation are now seen to be the result of relativistic motion on the atomic scale and of the consequential fields of time dilation, created by the presence of mass on the larger scale.

However, this leaves us with another problem. The concept of gravitational potential still presents the disturbing, unlimited reservoir of energy from nothing, ready to be acquired as kinetic energy, by any entity entering the gravitational field. This is still an apparent breach of the law of conservation of energy, so we need to understand the fundamentals somewhat better, in order to explain this.

The question is, whether the energy of the closed system is in balance at all times. "Is the energy of the closed system always conserved?" We must consider the energy of the closed system to include both the large mass of the gravitating body and the smaller mass being attracted in the gravitational field. We must also consider the time dilation at rest of both the large and small masses, since this is the attribute which gives any mass the potential to attract any other entity and is, therefore, a form of energy.

When the smaller mass is at a large distance, the system contains energy which approximates to the combined, static mass energy of both bodies plus the energy of the independent time dilation fields produced by each mass at rest.

$$E_{total} = Mc^2 + mc^2 + E(\Delta t_M{}') + E(\Delta t_m{}')$$

As the smaller mass "enters" the gravitational field and acquires kinetic energy, then the total system energy becomes the sum of the masses at rest + the time dilation of both masses at rest + the increase in kinetic energy of the smaller mass + the increase in time dilation of the smaller mass due to its speed.

It now becomes clear that, to balance the system, we have to regard the time dilation terms as negative and the kinetic time dilation energy as being equal to the kinetic energy. The energy of the system is then always conserved during free fall.

$$E_{total} = Mc^2 + mc^2 - E(\Delta t_M{}') - E(\Delta t_m{}') + \frac{1}{2}mv^2 - E(\Delta t_{KE}{}')$$

We can also deduce from the above, and from our proposed modifications to GR in chapter 6, that the mass energy of any body is equal and opposite to its static time dilation energy. The energy of all closed systems is then always a net zero.

When the smaller mass impacts the larger body, it transfers its kinetic energy into sound, heat, light, the destruction of the small body and the physical disturbance of the surface features of the large mass. So, the whole system always retains the kinetic energy of the smaller body, in one form or another. The total system energy is now the sum of the following: – The rest mass of both bodies – time dilation of both masses at rest + kinetic energy of the impact body – time dilation of the kinetic energy at impact.

This gives the same equation as before, except the numerical values of the kinetic energy and the time dilation, due to motion, are now larger and at their maxima, but are still equal and opposite. The mass energy and kinetic time dilation energy of the small mass finally exists as an increase in overall mass and static time dilation of the combined masses, plus any effect the collision has had on the orbit of the large body. The total energy of the system is, therefore, conserved at all stages.

The apparent conflict between the action of gravitation and the requirements of the law of conservation of energy is, therefore, now resolved. However, the nature of time dilation needs further philosophical investigation, since there is still an issue with the atomic relativistic effects producing unlimited energy on demand.

Nevertheless, it does now seem logical to conclude that the net energy of any closed system can now be regarded as always being zero.

This means the universe also, since the universe is the ultimate closed system.

We might therefore propose the following general formula for the total energy within any system, including the universe:

$$E_{total} = \sum \left[mc^2 - E(\Delta t_m{'}) + \frac{1}{2} mv^2 - E(\Delta t_{KE}{'}) \right] = 0$$

There is an implication in this that no net energy has ever been created, since any positive energy is always balanced by the consequential negative time dilation energy.

The Big Bang created nothing, but simply made some equal, plus and minus adjustments to the empty void. We will see a verification of this idea from the quantum scale up, later in Chapter 10.

This might also imply that gravitation must travel instantaneously, to prevent short term breaches of the law of conservation of energy.

Since this is not possible, relatively, due to the limiting speed of light, we must conclude that the total kinetic time dilation energy in the field is available instantaneously, as it is generated. But, it must be compressed within its smaller initial volume, as it spreads out at the speed of light, until the field has expanded to infinity.

In this way, the energy of the system is always conserved.

The spread of the field *is* instantaneous in the frame of the field at the relative velocity of "*c*", and so, as far as the energy of the field is concerned and as viewed from its moving frame, the spread is indeed instantaneous.

GR makes use of Pseudo tensors to deal with this energy conservation problem.

Orbital Motion and The Hammer and Feather

The new theory shows that there will always be acceleration towards the centre of the time dilatational field, without the need for an applied force. This acceleration is independent of the mass being accelerated, the acceleration being due entirely to the time curvature at any elevation in the field which, in turn, is related to the mass of the large object. This explains how and why satellites maintain their orbits at the same elevation, irrespective of their mass, and also why the classic hammer and feather accelerate to the ground at the same rate (in a vacuum). Observed behaviour has now been fully explained, but the issue can also be resolved by understanding better the equivalence of inertial mass and "gravitational mass".

Gravitational and Inertial Mass

The new theory conforms to the equivalence principle that gravitational acceleration is independent of the nature of the free-falling body.

The acceleration is proportional to the mass producing the gravitational field. We therefore, logically and independently, conclude that:

Gravitational mass = Inertial mass

Notwithstanding this, the new theory does not really require such an equivalence principle, since Newton's second law of motion is not invoked for an explanation of gravitational acceleration.

Einstein invents the concept of gravitational mass for the accelerating body, since he sees a non-inertial "force" (gravitational force), acting on a free falling body, yet he observes the acceleration to be independent of the mass in free fall. He has to explain why this is the case.

We do not need to explain the equivalence of inertial and gravitational mass, since we do not need to invent the concept of gravitational mass in the first place.

There is no such entity in reality, it is a fictitious property, a creation. This property was invented to make the mathematics work, but we can now see that the inertial mass of an object in free fall is simply irrelevant to the gravitational acceleration. The acceleration is due to the progressive dilation of the time term in the simple velocity equation,

$$v = \frac{dr}{dt}$$

and the very concept of gravitational mass becomes unnecessary.

The Bending of Light in A Gravitational Field

This is a most important issue and a critical one, as it turns out, for the new theory.

Acceleration due to gravity is now understood to be the result of the dilation of the time term, dt in the simple equations of motion below and not as a result of the application of any force.

$$v = \frac{dr}{dt} \quad \text{and} \quad a = \frac{d^2r}{dt^2},$$

Because of the nature of this cause of acceleration, we can actually "see" (it is obvious) that the effect will be independent of the mass of the gravitating entity. This not only means that bodies with different masses will accelerate at the same rate, but that, in the extreme, for entities with zero mass this will also be the case. Acceleration due to time dilation applies to everything that exists, be it a satellite, a meteorite, or a planet.

It is now clear that the gravitational acceleration induced by the time dilatational field, applies to all of these entities, including even a beam of light.

We no longer need to visualise light as particulate by nature, to allow us to understand how it is accelerated in a gravitational field and the problem with this aspect of the wave-particle duality of light transmission becomes irrelevant. Indeed, as an alternative to creating space-time curvatures to explain this effect where a beam of light follows a curved geodesic, we should consider a beam of light as simply experiencing the acceleration vector, due to time dilation.

This vector should be added locally and in respect of its position in the gravitational field, to its normal uninterrupted straight line path. This will describe the free fall Newtonian deflection of the beam.

Surprisingly perhaps, the observed deflection of the beam is *twice* that predicted by the formula for acceleration for pure Newtonian gravitation. Einstein predicted the correct, doubled deflection angle from his General Relativity theory (1915), and this was confirmed by the famous expeditions in 1919 to Brazil and West Africa to observe and measure these light deflections of starlight passing close to the Sun, under total solar eclipse conditions.

After these expeditions, Einstein made the following statement regarding the observed deflection of the light beams passing close to the Sun:

> *"It may be added that, according to the theory, (General Relativity), half of this deflection is produced by the Newtonian field of attraction of the Sun, and the other half by the geometrical modifications ("curvature") of space caused by the Sun."*
>
> *"Albert Einstein - Relativity", translation by Robert W. Lawson, Routledge Classics (ISBN10: 0-415-25384-5)*

This was interpreted to be a direct result of the General theory of Relativity of 1915, subsequent to the Special theory of Relativity of 1905.

The special theory uses Minkowski space, or flat space time. Flat space time does not give this doubling of the angle of deflection of the light beam, since there is only time curvature and no space curvature in flat space. If this were indeed the case, then the new theory would fail due to this one issue alone.

My instincts tell me that it would be a coincidence of great magnitude for a Gaussian space-time coordinate system which allows independent distortion in the four dimensions, to produce exactly the same amount of deflection from the distortion of three of these dimensions as that from the remaining one dimension (time).

When something exactly doubles in nature there is invariably a reason or a rule which gives this very specific result and coincidence to this degree never occurs in nature. In this case, it must be due to both the speed of the light beam being equal to "c", and the wave nature of light itself.

These are the only two attributes that set this event apart from the deflection of a solid object near the Sun, which has only half the deflection of the light beam, the Newtonian deflection, approximately.

Luckily for the new theory, there are others who disagree that the distortion of space is the cause of the other half of the deflection and it can now be demonstrated that Special Relativity does, in fact, make the correct prediction, without the physical distortion of space invoked by GR.

"...*If we take into account the wave nature of light, there is an <u>additional</u> contribution coming from the time dilation in relativity. The observation that this contribution is independent and additional to the Newtonian deflection of the mean trajectory is the main purpose of this paper.*

...One does not need full general relativity for deriving the expression for the deflection of light. What are needed are the equivalence principle, conservation of energy, and the wave nature of light. One part of the deflection comes from the free fall of the particle or light ray in the gravitational field, and the deflection depends on the average velocity of the test particle. The other part comes from the red shift factor. This is always given by, $2GM/c^2R$ independent of the velocity...

...For light, both contributions have the same magnitude, and they add to give the full deflection: $4GM/c^2R$ "

On the gravitational deflection of light and particles

C. S. Unnikrishnan

Current science, vol. 88 No. 7, 10 April 2005

Gravitation group, Tata Institute of fundamental research, Homi Bhabha Road, Mumbai 400 005, India.

This excellent paper, when read in full, shows why the curved space-time of General Relativity gives the same answer for the other half of the beam deflection as for the phenomenon of wave front bending. We now see that there are two simultaneous but independent effects occurring, both from one, common cause. The first is the Newtonian deflection due to the time dilatational field alone, and this is dependent on velocity. The second is due to the wave nature of light passing through the region of varying time dilation or *red shift* and this is dependent on the speed of light, but independent of the actual velocity of the entity. In the special case of a light ray, both deflections are equal, and the deflection is doubled.

To summarise, deflection is doubled by the summation of:

- Einstein – One part Newtonian plus one part curvature of space.

- Unnikrishnan – One part Newtonian plus one part wave front bending, both due to local time dilation.

The new theory survives this last serious challenge and indeed fits rather better than Einstein's explanation. Unnikrishnan is correct regarding the effects of time dilation on wave front bending and he also understands that the Newtonian deflection is also due to time dilation.

Einstein was philosophically wrong in his assertion that half the deflection is due to the distortion of space. He has had to create the abstract notion of the curvature of space and then the complex mathematics to work out, indirectly, the added deflection of the wave front bending caused by the *red shift*.

The new theory agrees with Unnikrishnan and views the Newtonian part of the deflection as being caused by the same time dilation that gives rise to the wave front bending. The real cause for the doubling of the deflection is, therefore, the same local time dilation, for both "independent" effects. There is, therefore, no conflict between observation, the predictions from General Relativity and those from the new theory. The curvature of space is not necessary to explain this effect. In fact, we are presented with a single *fundamental* cause for any change in velocity of any entity in a gravitational field, rather than there being two distinct causes, this is clearly preferred over double causation. This is also compliant with the Causality Principle and the guidance of Occam's razor.

Because of this explanation, General Relativity is again in question, since the cause of this doubled deflection of light, near a massive body, is now seen to be due to something other than the distortion of space. We have demonstrated that two of the key beliefs, in Special and General Relativity, are now seen to be unreal, namely the distortion of space to give the above deflection of light and the illusion of inertial length contraction.

If Einstein had understood that length contraction is an illusion and that this doubled deflection of light is purely due to time dilation, I query if he would ever have felt it appropriate to introduce the idea of the curvature of space and to apply the four-dimensional Gaussian coordinate system to General Relativity. Luckily for us, he was not aware of this and followed his clever idea of abstract curvature to complete the theory. GR is the prime example of a theory which gives the right answers, but which does not reflect reality. The reality is that a vacuum cannot be distorted. Einstein has created GR in such a way as to marry it with all other areas of science and, by doing so, has devised the space-time coordinate system and its associated mathematics, to always give accurate predictions. He has broken the mathematical trail and devised the rigorous mathematics which links together all known science and which is therefore guided and restricted by reality, or at least by proven rules of behaviour. Needless to say, this was a magnificent achievement and this new way of thinking about space-time in no way diminishes his work.

Some Rules About Time.

8.1   A frame of reference is defined solely by its time rate

8.2   For inertial frames, energy input is required to change the time rate from any constant value, in either direction.

8.3   In a gravitational field, energy is transferred from time energy to kinetic energy as a body falls. It will always lose time energy, never gain it.

The "geodesic" is the line of least action, the trajectory which involves the most efficient way for a body to lose time energy. . It is the path that ensures the next position along it, always has the slowest clock.

8.4   Any physical entity, (a body or electromagnetic radiation), will always follow a path, ("gravitate") from a position with a faster time rate to a position with a slower time rate, without any force or energy being applied. Slower clocks "suck".

8.5   Newtonian gravitational acceleration (g) is proportional to the time dilation factor at any particular elevation in the gravitational field.

8.6   The time rate in all frames of reference must always be between the finite limits of universal time, t_f and zero.

8.7   More than one time-dilation field may coexist in the same location and the effective, resultant field is the vector sum of the fields.

8.8   Where more than one time-dilation field exist in the same location, they do not cancel each other, but coexist and counteract or accumulate at any point in the fields, depending on their relative directions at that point. The energies of both fields still exist.

8.9   If two frames of reference have the same time rate then, no matter how they have achieved this, either by speed or the elevation in a gravitational field, or by a combination of both, they are, in fact, the same frame of reference. (See 8.1).

Edwin Hubble's Mistake

Georges Lemaitre, a Belgian catholic priest (and a physicist first), predicted the increasing *red shift* of distant galaxies two years before Hubble. However, Hubble has been credited with the discovery since he observed the phenomenon whilst Lemaitre had merely predicted it, theoretically, from Einstein's field equations.

This increasing *red shift* with distance has been extrapolated, back in time, to predict the beginning of the universe from an infinitely dense singularity which gave rise to the Big Bang. The "Big Bang" is currently the clear favourite as a theory of creation, perhaps to such an extent that any alternative ideas may not be given due consideration. The physical expansion of the universe was deduced from this observed *red shift* of distant galaxies, because *red shift* is usually caused by the Doppler Effect. An example of this is the pitch of the whistle of a train as it passes us by and then lowers in frequency as the train starts to draw away from us again. The sound, which is of a wave form, is "stretched" into a longer wavelength by the receding train and so has a consequently lower pitch or frequency. Light, also being of a wave nature, is also known to exhibit this Doppler Effect.

The new theory predicts a different causality for this observed *red shift* that is not fundamentally, produced by physical expansion. The galaxies are not moving away from us for no reason, but because their time rates are slower than ours. Mind you, as can be seen from Chapter 7 on General Relativity, an increase in time rate also gives rise to the effective expansion of space, so we can be forgiven for interpreting this *red shift* as a Doppler effect. It amounts to the same thing, observationally. Time rate change is the only phenomenon that can sensibly explain the expansion of space, whilst Dark Energy is our invention of unknown origin, so I deduce that dark energy *is* the increase in the rate of time over time.

Dark Energy is the accelerating rate of time

The observed red shift is the same as looking into a gravitational field with a slower clock than ours. The reason for the red shift is not that we are looking into a gravitational field, it is because we are looking into a region with a slower time rate than ours and we will see the same effect when looking into the past, when time ran slower back then.

We call it *red shift*, since red light is at the long wave length or low frequency end of the visible spectrum. When Hubble observed that the complete light spectrum of distant galaxies is *red shifted* (wherever he looked in the heavens), he concluded that these galaxies are expanding away from us and that the universe, in general, is expanding as a whole. He then extrapolated back in time and assumed that there must have been a Big Bang to start the expansion off from a single point in the beginning.

This idea fails, unless we invent the concept of the expansion of space itself, but, to explain the fact that the *red shift* is the same looking in any direction, we must argue that space itself expands, like a loaf of bread rising in the oven. Any seeds within the loaf all move away from each other, wherever they are in the loaf. This explanation is an attempt to stop the observer having to be at the centre of the universe. Again, the major flaw in this reasoning is the idea that nothing (space) can actually expand at all and this is of course, philosophical nonsense, unless we can identify the causality for this effect.

The temporal (and more fundamental) cause of galactic *red shift* has not been considered as a serious possibility. This, I believe, was Hubble's duty to identify and explore, but he missed it. This alternative is the stretching or compressing of the wave form of light, due to a change in the universal time rate over the aeons. There is no scientific imperative for the assumption that the rate of passage of time (and the associated speed of light), has always been at the exact same rate, relatively over the life of the universe. We should therefore consider how the slowing, or quickening of time, might create this observed *red shift*.

If, in our universe, time were slowing down, then time would appear to be faster in the past than it is now and a clock at the point and time of the emission of light would have a faster time rate relative to a present day clock when the light is received. In the case where time is slowing, the observed light would in fact be *blue shifted*, and this would be due to our looking outwards into a region with a faster clock. So, we must conclude from the observed *red shift* that time is *not* slowing down but is *speeding up!*

I first proposed this idea early in 2011 in my book, "Time Dilation the reality", the forerunner to the first edition of "The Binary Universe". Recent articles have now acknowledged this possibility of variable time and suggest that time is slowing down but this idea is wrong. Since we observe *red shift* from light transmissions from the past, the clock must have been have running slower then and it now runs faster than at the time of the light emission.

So, time passed more slowly in the distant past than it does now, and if we extrapolate backwards again we are facing the prospect of a Temporal Eruption, rather than the physical Big Bang we all know and love. It seems the beginning of the universe was actually the beginning of time. In Chapter Twelve, we will see that this is the only sensible conclusion to draw and that the universe is slowly losing energy like a spinning coin, speeding up in frequency, as the spin runs down. So, if we assume the simplest explanation, then the *red shift* of distant galaxies is showing us that the universe is slowly and imperceptibly *speeding up* in time and that the expansion of space is an emergent effect from this changing time rate.

We have shown, in Chapter 7, that the time rate, effectively, defines volume. Slow down time and space compresses, but speed time up and space "expands", relatively. Speeding up time, or expanding space, is the equivalent to moving outwards in a gravitational field.

To anyone living in any of those space-times, the changing time rate does not matter since all processes will be speeding up. The laws of physics are still the same at any moment, whenever that moment is, now or a billion years ago. It is only when you make comparisons between past and present that you will observe that time was slower in the past and more so the further into the past you look. The *red shift* will be greater, the further away the observed galaxy is.

Furthermore, the fundamental problem with the Big Bang theory is not shared by this temporal explanation. If the universe is expanding physically in all directions then, as explained, we have to invent the inexplicable expansion of space, itself, to allow the Earth *not* to be at the centre of the universe. The expansion of nothing, without cause, is clearly a ridiculous notion but, if you have an increase in time rate, then the expansion of space is an inevitable result from this phenomenon. Both aspects are true, but the time rate change is the more fundamental.

Being dependent only on time differential and therefore on distance, the temporal shift will, naturally, be equal in all directions, from wherever the observation is made.

The degree of shift is dependent only on how far back in time the light was emitted and therefore upon the distance of the galaxy, not on the location of the observer, nor on the direction of his observation. I suggest that this postulate is a far more logical and, therefore, a much more likely proposition than simply space "expanding", without identifying the Dark Energy needed to produce this effect.

I believe that Edwin Hubble's creative, yet dubious, Doppler explanation is incomplete.

Another clue as to the likelihood of this proposition is the Hubble constant, the number that represents this *red shift* or physical expansion of the universe.

The Hubble Constant

We are proposing that Hubble's universal "expansion" is the changing of the time rate over time and so predictably, the Hubble constant has units of inverse time.

The Hubble constant is:

$$H_o = 2.3 \times 10^{-18} \, s^{-1}$$

Hubble time, the inverse of this is:

$$H_t = 4.35 \times 10^{17} \, s$$

This is a glaringly obvious demonstration that the universal "expansion" we observe is, more fundamentally, the quickening of time. It is the minute changing of the rate of time, between the ages, that gives rise to this *apparent* Doppler *red shift*. More evidence for this effect is shown by the phenomenon of Dark Energy.

Dark Energy

Dark Energy is demonstrated by the cosmological constant, "Λ" (Lambda) which has a tiny value of 10^{-122} in natural units. I am proposing that vacuum energy or dark energy is simply due to this quickening of time. It only becomes significant at very large scales, or over long time periods, since it is a tiny effect over time.

Crucially, if the universe were only physically expanding, then its mass would become more spread out and gravitational attraction would decrease as a result. This is a persistent conundrum, in today's science, that the energy density of space remains constant whilst it expands, an impossible situation.

Since the effect of an increasing time rate is the same as that from moving away from a mass, then it is understandable that we have misinterpreted its effects as pure physical expansion. Nevertheless, we have ignored the possibility of a changing time rate and the fact that this could also explain our observations. This idea should have been given equal priority to physical expansion, before one idea was then preferred for specific reasons, but it seems we have not even considered the possibility. If we truly have an open mind, we must consider the time rate changing over time, as well as a physical expansion, or even contraction, but to entertain only one of these options is, frankly, unscientific.

Our "creation" of Dark Energy is a hasty assumption to try and explain the energy in the vacuum throughout the universe, the increase in the rate of expansion. We should have also explored the possibility of Dark Energy, as the energy equivalent of the changing rate of time over the eons.

The Hubble constant is $H_o = 2.3 \times 10^{-18} s^{-1}$ and in natural units is:

$$\frac{1}{1.855 \times 10^{43}} \times 2.3 \times 10^{-18} = 10^{-61} \, Plancktime^{-1}$$

And so,

$$H_o^{\,2} = 10^{-122} = \Lambda = \frac{1}{T_H^{\,2}}$$

So, what this means is,

> *Dark Energy = The square of the Hubble constant*

or

> *Dark Energy = The inverse of the square of Hubble time*

Since the Hubble constant is the inverse of Hubble time, and I am proposing that the expansion of the universe is a misinterpretation of the changing of universal time, then this is a compelling argument that dark energy is simply the energy due to the changing time rate over the eons. It might appear that universal expansion is accelerating, but it is the time rate that is speeding up and space is simply "expanding" as the emergent effect. If my conjecture is correct and time is *the* fundamental, with space emerging from time, then the driving force for "universal expansion" is in fact an increasing time rate.

Dark Energy = an acceleration of 5.29×10^{-36} $(/s^2)$

In natural units:

Dark Energy = an acceleration of 10^{-122} $(Plancktime^{-2})$

Note that the units are of inverse time squared. This is a clear indication that time *is* energy!

This is a very small value, but the effect exists throughout the whole volume of universal space, so the total energy is significant. In fact, it is estimated that dark energy is around 70% of the total energy in the universe, excluding normal matter which constitutes only about 5%. Most of the universe is empty space, but it all contains this miniscule energy from the speeding up of time.

Later, we shall see why this vacuum energy is so low, compared to the predictions from quantum field theory of a much larger number and how that is almost completely cancelled out.

9 The Experiment

This theory will stand or fall, depending on the results of my proposed experiment to measure the *blue* or *red shift* of light, when observing the less accelerated frame from the faster moving one.

Transverse Shift

We have mentioned the *blue* and *red shift* effects observed when looking outward from and inward to, a source of gravitation respectively.

These effects are predicted by accepted theory and confirmed by observation. Contrary to Special Relativity, the new understanding of time predicts that the same *red* and *blue shift* effects will be observed from the stationary frame and the moving frame respectively, but independently from Doppler effects. The frequency shifting for both gravitational and inertial systems is due to the time rate differential between frames. In the case of inertial systems, the shifting is caused by motion, instead of by a gravitational field, but you cannot have different types of time dilation.

Time is either dilated, or it is not, and the observations of one frame from the other must give the same results, independently of whether the time dilation is gravitationally generated, or generated by motion.

A clear understanding of transverse shift has been thwarted by the attempts to explain it away using Doppler shift. This is an example of a belief being forced onto science, despite the clear evidence to the contrary.

Any inertial time rate differential between frames depends on the common frame of reference that the two observers shared initially and the difference, in acceleration, between the two.

This is not a Doppler effect with the wavelength and frequency changing due to relative speed, but is due to the time rate differential between the different frames, the transverse shift. It is caused by the greater time dilation of the more accelerated clock. My *blue shift* prediction on observing the stationary frame from the moving frame is due to the non-reciprocity of time dilation and is counter to the reciprocal *red shift* prediction from Special Relativity.

The reciprocal *red shift* prediction from SR has still to be experimentally tested and it will fail such a test. The proposed test should be carried out with two-way observations between one fast moving space vehicle and one slower vehicle or "stationary" location. This will confirm that *blue shift* is seen when observing "stationary", or less accelerated, light signals of known frequencies, from the fast test bed moving parallel to the other.

Critically, the observation must be done throughout the closing and separating phases (approach and divergence), including the moment of their passing. This test will place high demands on the sensitivity of the instrumentation used and on the detailed specification and methodology of the experiment.

Figure 17 – shows the frequency changes due to Doppler shift only. I have left out any allowance for inertial time dilation and its associated frequency shifts, to set the baseline for the test. This is analogous to the whistle of a train changing from when it approaches to when it moves off into the distance.

The graph represents the tone of the whistle, high on approach, dropping as it passes, then lower as it draws away, but in the case of our test it is the frequency of light that the graph is showing, not the frequency of sound.

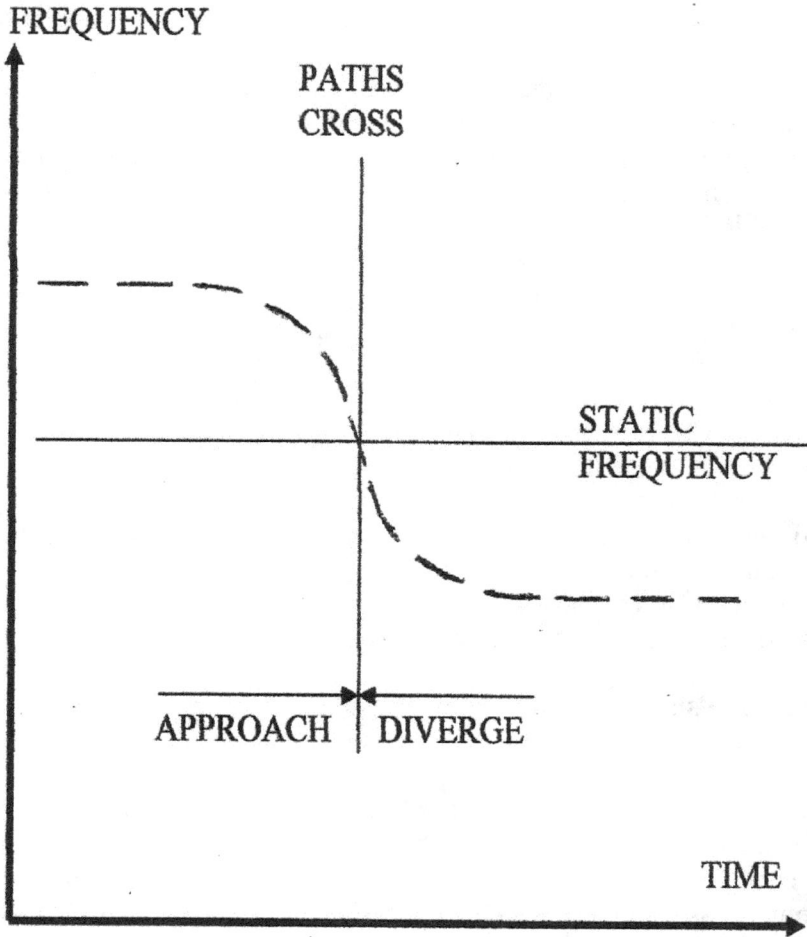

Figure 17: Pure Doppler shift without inertial time dilation

Observing the fast source from the 'frame of the less accelerated source, SR predicts an additional *red shift* over the whole range of the graph, due to inertial time dilation. The onboard time rate is known to slow down, due to its motion in accordance with the Lorentz transformation. So, the whole graph actually drops down by this transverse, inertial *red shift*, from the position of the pure Doppler shift curve. This observation from the slow test bed is shown in **Figure 18.**

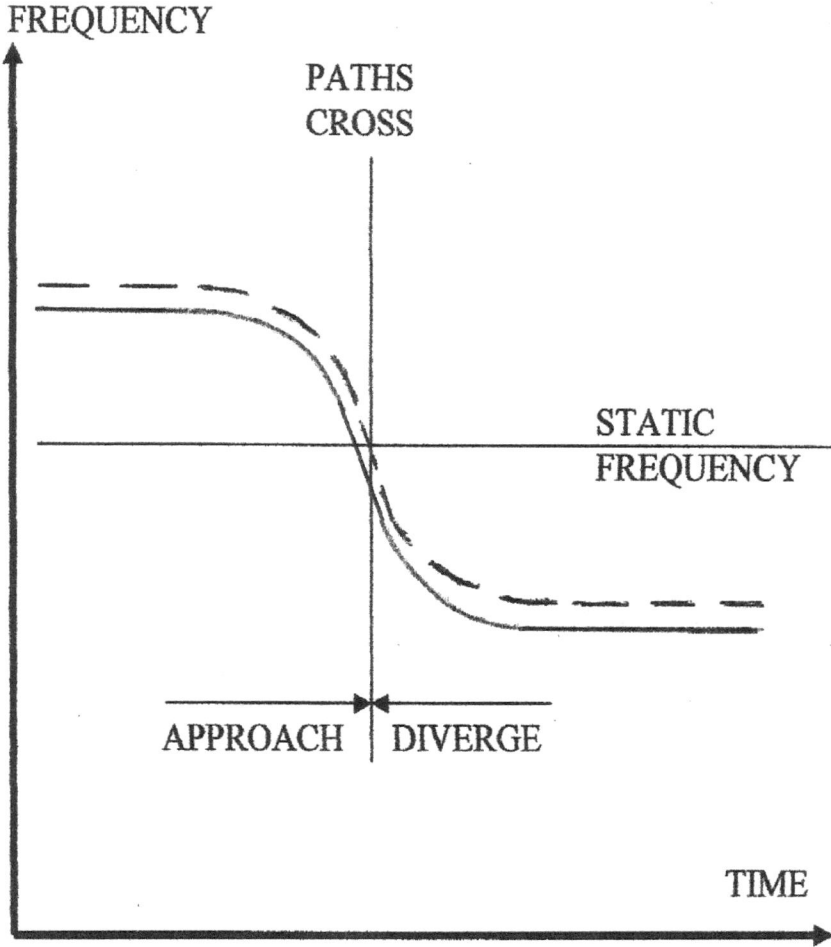

Figure 18: Transverse *red shift* observed in the accelerated frame

Now, SR predicts that the observation from the fast test bed, looking at the slow test bed, will also show this same, *red shift*. SR predicts the same graph as in **Figure 18** from both perspectives.

This *opinion*, and it is only an opinion rather than some mathematical imperative, stems from a certain fear in mainstream thinking.

They *choose* to believe that all motion *has* to be purely relative, so that they do not have to entertain a preferred reference frame against which all motion must be measured.

In doing this, they have declared the result before testing it and have written out any other possibility from their plans.

On the other hand, I am not afraid of any repercussions from following obvious and simple logic.

Since we are now observing the slow/stationary test bed which has a faster clock than the faster moving test bed, then we must observe a transverse *blue shift* of the light from the slow-moving test bed. This *blue shift* prediction is counter to predictions from SR due to our difference in opinion.

By making this choice, I am committing to defining the preferred reference frame. I will do this in the next section and I will clarify in more detail later on, when discussing our Binary Universe. The mainstream cannot bear this heretic idea,. To accept it means that their version of physical reality is destroyed. But, the reality of the preferred reference frame finally explains the unresolved problems of "Newton's Bucket[53]" and the myth of Mach's Principle[54].

We now have space vehicles which can travel fast enough to exhibit measurable inertial time dilation and associated *red shift*, as well as instrumentation sensitive enough to detect it and so this experiment has now become a practical proposition.

---

[53] Newton's bucket – The problem of explaining how water in a rotating bucket "knows" that the bucket is rotating.
[54] Mach's Principle – Explains Newton's Bucket by reference to the fixed stars in the heavens.

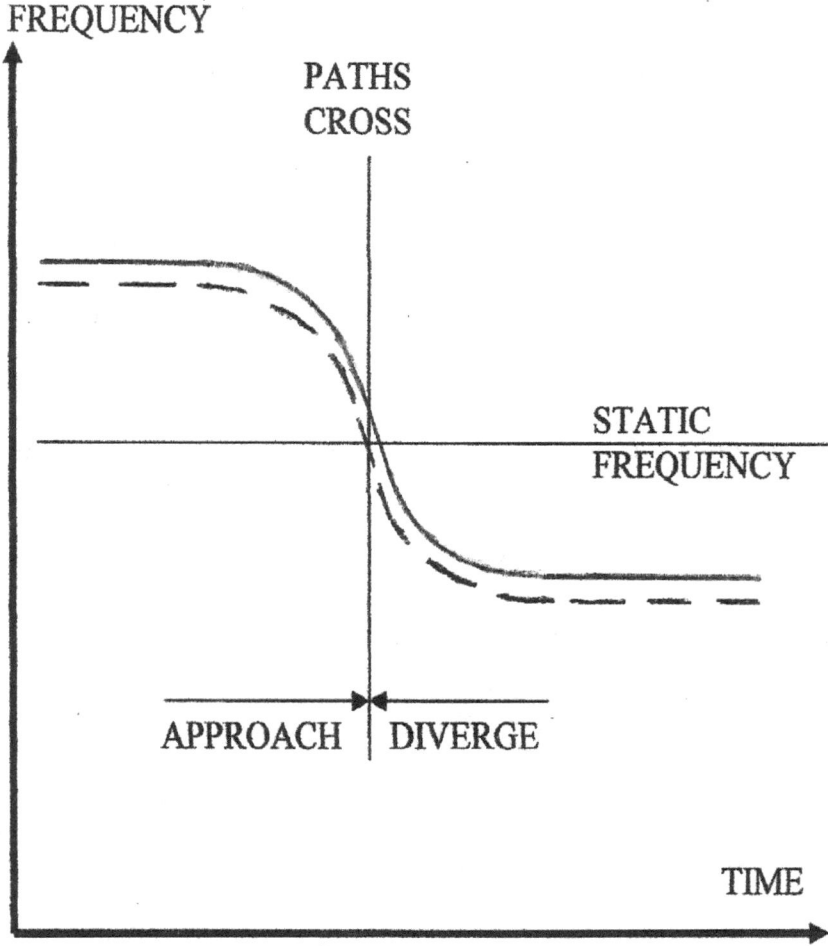

Figure 19: Slower frame *Blue shift* observed from the accelerated frame

Figure 19 shows my prediction of a *blue shifted* light observed from the more accelerated frame. The frequency is higher at all positions over the curve.

There is no excuse for not integrating this experiment into some future planned space mission and resolving this long-standing issue once and for all. The complete theory, as proposed, will stand or fall on the results from this test. Positive results will demand immediate acceptance of these new ideas and a revision to our understanding of Special and General Relativity. Clearly, the implications are far-reaching.

The Preferred Reference Field

The definition of a field is that the value of the effect changes with position in the field. A magnetic field has different values of magnetic effect as you move through it. An electric field has the same varying effect with position, but this time with electrical charge. So, if the time rate varies with position in space, and we know that it does throughout all gravitational fields, then we can say there is a time rate field throughout space. The time rate field is not static. It is more a swirling field of differing time rates formed by all the moving, overlapping, spherical, gravitational fields in the universe. It is irrelevant that it is dynamic. It is still a frame of reference locally, despite an infinite number of frames of reference throughout space, ranging from zero at the event horizons of black holes, to almost universal time in the far reaches of interstellar space.

We have established, earlier in Chapter 8, that a frame of reference is defined purely by its time rate, so the local time rate field is the preferred reference frame. It is the aggregate of all the effects from all gravitational fields against which we must compare all motion and events, from the cosmic scale down towards the Planck scale. The local time rate field is the local preferred reference frame.

Mainstream science claims there can be no preferred reference frame since all velocity is deemed purely relative, but we are simply *told* to believe this without any evidence to support it. The mainstream' version of SR is compelled to believe that *red shift* is observed from both the less accelerated and the more accelerated frames', since this is an essential element in the belief in all motion being purely relative. They believe in this symmetry of observation since they claim that it does not matter which entity moves faster than the other and it only matters that the velocity of each looks the same from the other's perspective. If this really were the case, then the Hafele and Keating clocks would have been unaware of the different accelerations between the ground based and airborne clocks, but they clearly "knew" about this difference. They experienced it and recorded it.

If the mainstream were to accept the real asymmetry of time dilation, as predicted by Albert Einstein and proven by Hafele & Keating, then they would be forced to accept that both entities must move relative to some other (preferred) frame at different rates and that this, preferred reference frame must be the field that produces inertial time dilation, as you move through it. This is the time rate field. The preferred reference frame, is the "taboo" of traditional relativity. The very idea is sheer heresy to today's mainstream physicists, but it is, nevertheless, correct and I sincerely hope it does not take another century for it to become understood and accepted.

I think Newton had the right idea with his "Absolute Space" but of course, he was unaware of the exact nature of time and of the phenomenon of inertial time dilation, so he could not explain it, fully.

Newton's Bucket and Mach's Principle

The issue of Newton's bucket is the famous debate over whether motion is purely relative or absolute, against some preferred reference frame. In his thought experiment, Newton presents the situation of a bucket, half filled with water and suspended from a wound-up string, or rope, which is then allowed to unwind. The bucket starts to rotate and, at first, the surface of the water is flat, but as the fluid friction between the water and the rotating bucket walls takes effect, the water eventually gets dragged around by the bucket walls. The water's surface has now become concave, due to its rotation about its central axis. Newton argues that the water did not care at first about the relative rotation of the bucket, so relative motion is not important. The water only notices it is rotating when it does so against some other, invisible frame of "absolute space", his preferred reference frame. Newton therefore argues that all motion is absolute, not purely relative.

Mach's principle, initiated by Ernst Mach and invoked by Einstein in his logic for general relativity, is counter to Newton's bucket argument. It uses the distant stars to show how the relative rotation of the stars, as the water rotates in the bucket, defines the water's rotation. He argues that the surrounding mass of everything, including the distant stars, determines whether or not the water rotates.

Although these ideas are flawed, there is much truth in both of them. Newton was right to suspect that relative motion is irrelevant to the water, but, understandably, he failed to identify what his "absolute space" is. Mach is right to suspect that this relative motion is, in some way, connected to the fixed stars, but he fails to demonstrate how the fixed stars can affect local events. Both Newton's and Mach's explanations are, therefore, incomplete. Let me complete both of their arguments and we will see that they both converge on the one idea, the idea of the *standing* wave of time, the time rate field.

As is always the case, the key to understanding any interaction is causality. For every effect, there must be a proximate, antecedent cause. The cause must be local to the effect and must pre-exist the event. Only a field can do this. We have already shown that the gravitational field is just the time rate field and we have demonstrated how GR can be restructured to accommodate this idea, by removing the abstraction of the curvature of space and replacing it with the real changes in time rate, the changing energy of time.

What we are left with, is the time rate field, the field that exists throughout the universe, since time passes at some rate or other everywhere. The time rate field is proximate to everything, including Newton's bucket (which sits in the field before, during and after its rotation). The field is always proximate and pre-existing to every event, anywhere. So, by elimination only, this field is the only entity which can cause the water to form its concavity when rotating. In fact, it will become more obvious, later, that this field is ultimately the cause for *every* effect above the quantum domain.

In Chapter 8, rule 8.4 states that any physical entity will always try and move towards a position with a slower time rate. This is also true of rotating masses and is the underlying causality for the Earth being oblate, or egg shaped. When a body of water in a bucket rotates about its central axis, there is a tiny degree of inertial time dilation which increases from the centre towards the perimeter of the bucket. Rule 8.4 means that all the water in the bucket will tend towards the perimeter, where time runs a little slower. As an Engineer, I am aware of the mathematical techniques of centripetal acceleration, but these are only techniques. Time is always the fundamental reason for everything. This effect increases from zero, at the centre, to a value relating to the time dilation from the peripheral speed of rotation, at the extremity. The water tries to pile up towards the outside of the bucket, where time runs slowest. This is the fundamental causality for the centrifugal behaviour, not some absolute space, nor some relationship with the distant stars. Newton's absolute space does contain time, of which he was unaware, so he was unable to define his absolute frame of reference. All mass, including the distant stars, will affect local time, so there is some connection, if only miniscule, with local events. But the distant stars are not the cause of local events. All that the distant stars do for us, is to demonstrate the non-rotating frame, which if we bring that frame near to us, becomes the local, non-rotating frame and this frame still requires to be defined. There is only one way to define this and that is by the time rate field. So, Newton and Mach were on the right track but failed to complete their arguments because they were unaware of the concept of the time rate field and of inertial time dilation. It cannot be over emphasised that every real effect requires a real, proximate, antecedent cause and if we always hang on to this golden rule, we will always be able to understand things better.

The Field

You can envision this preferred reference field and "see" that it must be true. At some relative "speed", in intergalactic space, you will be stationary! Yes, absolutely stationary! You will be at a speed where all the light you are receiving, from all directions, is neither *red* nor *blue* shift*ed* by your motion. The spectra are all perfect from every direction. Of course, this supposes that all distant light emitters have a common velocity relative to you, so we would have to take the average spread of light shift from all directions and assume a certain homogeneity of motion throughout the cosmos. Even the *red shift* from the "expansion" of the universe is experienced as the light from all directions having the same degree of *red shift*, so this also has no effect on direction.

From this absolutely stationary state, any movement from this position, in any direction, at any speed will *blue shift* all the light ahead of you and *red shift* all the light behind you. In fact, this observation has already been made, although it is not fully recognised for what it is, but is well known as the "Dipole Anisotropy" which is indeed put down to the speed of the Earth through local space.

In your absolutely stationary position, your clock will run at almost "Universal time", the fastest time rate possible, assuming you are at great distance from all mass. As soon as you move at any speed, in any direction, your time rate slows down from this maximum time rate. You can now see that there really *is* a particular speed at which this happens. This speed is the absolute zero speed. It is also coincident with a zero Doppler shift in both the fore' and aft' directions, in fact, in all directions.

The dynamic field of the standing wave of time is the preferred reference frame.

The speed of the Solar system through intergalactic space is currently estimated to be about 635 km/s, in a direction towards "The Great Attractor", not exactly light speed but fast enough to produce the observed, detectable Doppler shifts of the light. If the time rate field, the *standing* wave of time, *is* the preferred reference frame (and it is clear to me that it must be), then everything must move relative to this field. In this case, we must revert to the idea of the Earth moving through what is, essentially, the new "Luminiferous Aether"! The speed of the Earth, through this field, is not an appreciable fraction of the speed of light and we are drifting through the time rate field at some small velocity (635 km/s). Consequently, our time rate is already slightly dilated, in absolute terms, due to this speed, as well as from gravitational effects. But the speed vector of any point on the Earth varies in magnitude and direction throughout the year, even during the day, so the inertial time dilation, due to this motion, at any point on the Earth, will also vary with time.

The Earth's absolute speed must be envisioned at the intergalactic scale. Our galaxy is moving through the field of time at a finite velocity, in just one direction. Our Solar system is rotating around the galactic centre and, on top of this, our planet is rotating within the Solar system, as well as spinning about its own axis. This means the movement of any point on the surface of the Earth, through the field of time, varies in magnitude and possibly direction as our planet spins and revolves around the Sun, as we rotate around our galactic centre and as our galaxy moves through the intergalactic field. The absolute motion is complex and difficult to unravel, but not impossible.

Firstly, we might measure any small, unpredicted *red* and/or *blue shifts* of probe signals travelling in different directions and the Flyby Anomaly[55] might be a good initial assessment of this overall effect. The Pioneer Anomaly might also display this effect. In order to make sense of the unpredicted shifts which have been observed from various probes at perigee[56] (as they flew by the Earth), we need to know the location of the perigees above the Earth's surface, their angle of inclination from the Earth's orbital plane, the time of day they occurred, the dates they occurred (positions around the Sun) and the speed and direction of the solar system around the galactic centre. We should then be able to make an assessment of the absolute speeds and directions of all the probes through intergalactic space at the time of perigee, to see if there is any correlation with the observed shifts in signals. We already have the relative speed vectors and Doppler shifts for various probes and, with all this data, we can resolve the vector components to see where the maximum and minimum shifts occur and in which absolute directions.

The unexplained *blue shift,* observed from the pioneer 1 and 2 probes, could also be analysed to see if there is any correlation between the speeds and directions of the probes and the direction for the Dipole Anisotropy. We should expect to see a slight *blue shift* of light in the direction we are travelling through the Aether and an equal and opposite *red shift* in the opposite direction, behind us. The new theory predicts this general "patch" of slight *blue shift* and a general "patch" of slight *red shift* in the exact opposite direction. In other words, the theory predicts the "Dipole Anisotropy", which is already compelling evidence of the time rate field. If this also matches up with the results from the FlyBy and Pioneer analyses, then this would be undeniable evidence of the existence of the standing field of time, the new Luminiferous Aether, (should we wish to call it that).

Finally, we must realise that the Dipole' Doppler shifts, on their own, do not *prove* the existence of the preferred reference frame,

---

[55] Flyby Anomaly – Observed, unpredicted red and *blue shifts* of signals from various probes at perigee near the Earth.

[56] Perigee – The point at which an orbiting moon or satellite is nearest to the Earth during its elliptical orbit or parabolic trajectory.

although they are a very strong indication of it. The Doppler shifting is one of the *two* things that do so, together. Firstly, the stationary state is the state where there are no Doppler shifts ahead, or behind. It is also the state that shows us the maximum transverse *red shift* of everything, in all directions, since our clock is running, at the fastest possible rate. These two effects are coincident.

Exactly when we might be able to carry out this experiment, in intergalactic space, I do not know, but it will certainly not be soon.

10 Binary Universe Theory (B.U.T.)

Well, here it is at last, the final payoff for your perseverance and for having read thus far. Up until now, I have merely exposed our mistaken beliefs in science, but now it is time to analyse, fully, the implications and to explore the new reality emerging from the resolution of these errors.

The "Arrow of Time"[57] grows at the rate of 1.855×10^{43} Planck time units every second, today, it is approximately 13.8 billion years long. The universe exists only on the very tip of this arrow, in the ever changing, yet always immediate, present. Looking back down the arrow's shaft, we see all history, which is certain and fixed. Ahead of the arrow's tip is the uncertain future.

There is still much philosophical discussion over the nature of time, including whether time is subjective or objective, but I do not propose to enter into these fruitless debates. Time is an objective, physical phenomenon and this chapter will prove this beyond doubt. The philosophical debate as to the nature of time will end, here.

This chapter will show:
- *how* and *why* light is of a wave nature and *how* and *why* time slows down with speed.
- It will identify the direct cause and effect for both phenomena.
- It will also explain the wave-particle duality of light.

Surprisingly, present day science has no satisfactory explanation for any of these effects.

Additionally, it will be demonstrated that there *is* a preferred reference frame, or more accurately, a preferred reference *field*.

---

[57] "Arrow of Time" – A simple way of understanding the progression of time with length and direction analogous to duration and the future.

We will be driven to the ultimate conclusion that we are living in a Binary Universe and in Chapter ll, we will test this postulate against many aspects of accepted science and against certain present day conundrums.

We will see that the B.U.T. stands the test of scrutiny against all of these, perfectly. These ideas may be shocking to some modern physicists, but the arguments presented and the resulting conclusions are compelling.

When we carefully consider the implications of the proposed modifications to Special Relativity, we begin to see more clearly that the speed of light is directly related to "universal time". The speed of light is the particular relative velocity at which universal time has been dilated to zero, for the traveller (photon). This particular value of "c" in the geometric proof of time dilation and in the Lorenz factor is directly related to the degree of time dilation in the moving frame. It is the value of relative velocity, at which the velocity experienced in the moving frame becomes infinite. If we were to change this natural limit of "c", we would change these relationships, but we would be in a different universe, with different rules.

Here we are going to consider the rate of passage of time. If we understand and accept that time passes slower, deeper in gravitational fields and also at high speeds, then we have to accept the reality of different time rates within different frames of reference and at different locations. We have to acknowledge the variable time rate field and the fact that time passes at a certain rate, relatively, at any particular point in space. As a result, we must also accept that the time rate would revert to "universal time" in the absence of gravity and of motion through the field of time.

Universal time is the fastest time rate possible. It is the theoretical rate of passage of time in a theoretical space where there is no gravitational field. There may be nowhere in our universe where this time rate occurs, since there is always some residual gravitational effect and corresponding time dilation, but universal time is the time rate at which all "stationary" clocks naturally run, without interference from the presence of mass.

It is said that "The speed of light *is* the speed of time" and this is quite true in terms of energy. With zero kinetic energy, our time is not dilated, it runs at "full speed", at maximum temporal energy. At the speed of light, our time has stopped altogether and does not run at all, but our motion is now at "full speed", at maximum kinetic energy. Clearly, the energies of time (and of speed) transfer between each other as speed changes. This means that time *must* be a form of energy. We shall prove that time is energy and that the energy of time does transfer to the energy of motion with increasing speed, then transfers back again, as we slow down. The first question is, does this time energy occur as a smooth, continuous flow, or in discrete units? Is time continuous or is it quantised? Does time "flow" or "jump" forward into the future?

No analogue (non-digital) process is completely free from interference, nor can it be completely accurate. Digital processes, however, are completely free from interference and are also completely accurate. We know this from our experiences with digital radio and television transmissions, as well as digital computers and other equipment. For the phenomenon of time to be free from error, free from degradation and to be completely accurate, it is inconceivable for it to be an analogue process. It must be a digital process.

It is also noteworthy that nothing, no process, can even start without breaking the rules, if time is a continuous or analogue process. As we accelerate from "rest", or indeed from any speed, there is the problem of getting from zero energy to some positive value and theoretically that requires infinite "jerk"[58], a momentary, infinite rate of change of acceleration with an infinite spike of energy. Only with a quantised time, can this energy be transferred to kinetic energy avoiding this problem. The energy of one-time quantum is finite. The instantaneous energy required to start motion can never be greater than this amount.

If time were actually a continuous "stream" of something, then there would be no reasonable process by which we could make different amounts of temporal progress, to experience a different number of changes, unless time were quantised. We are led, therefore, to the notion of a quantum time, the quantum being the Planck time, the minimum period between any two events, or minimum duration of any action. There is a finite number of

---

[58] Jerk – the rate of change of acceleration, or the derivative of acceleration with respect to time.

changes available in the form of 1.855×10^{43} Planck times per second of universal time, but we are not all compelled to use these at the same rate for temporal progress. Indeed, we are compelled to use them at a slower rate to move through time, when we use some of them for things other than for temporal progress. This depends on our frame of reference, as we shall see later.

Therefore, from these aspects of nature, I deduce that time *must* be quantised. It must pass in discrete units of time or in quanta of the energy of time, namely the Planck time.

Quantised Time

At the finite speed of light, the maximum speed attainable, you travel one Planck length in one Planck time. This is the definition of these "natural" units. The Planck time is considered to be the smallest possible increment of time. Some scientists now take the view that time can run quicker than this unit but I fail to see how. The Planck time is the time it takes light to travel one Planck length at the speed of light. If we fix the Planck length as the fundamental unit of space then at the maximum speed of "c", it takes one Planck time to cross it. Alternatively, we could fix the Planck time and let that define the minimum unit of length. Whichever choice we make, is a matter if opinion. Since we have seen that space emerges from time and that time is *the* fundamental, then it is imperative we identify the unit of time as being fundamental and not the unit of space. Even if we were to fix the Planck length as a fundamental, then the quickest time to cross it is still one Planck time. We can only slow down from light speed, not speed up, so for slower speeds than "c" it will take longer than one Planck time to cross a Planck length. So, it must be possible to travel fractions of a Planck length, but you cannot proceed into the future in less than one Planck time.

The Planck length is *not* the minimum distance you can travel, it is the *maximum* distance you can travel in one Planck time. For slow velocities, clearly you will experience more than one Planck time to cover one Planck length, so the Planck length is divisible by the number of Planck times used to cross it. In the case of a very slow-moving entity which takes, say, one second to cross a Planck length, then the number of Planck times taken to traverse it approaches 1.855×10^{43}, so the Planck length can be divisible by at least this number. Clearly the Planck length is not a fundamental unit of length, but the Planck time is *the* fundamental unit of time, in fact, it is *the* fundamental unit, period! This is because time is quantised, whereas the void is continuous.

Scientists today cannot detect, or measure, effects anywhere near this very small/fast scale. One Planck length is 1.6162×10^{-35} metres, which is about 10^{-20} of the diameter of a proton. The unit of Planck time is a mere 5.391×10^{-44} seconds and, as of November 2016, the smallest time interval directly measured was 850 zeptoseconds or 850×10^{-21} seconds.

There will be smaller measurements, of course, as our experimental techniques improve, but we will never be able to measure time increments anywhere near the order of magnitude of the Planck time. The definition of these natural units comes from a process called "dimensional analysis" and they stem from the fact that, when you travel at the speed of light, you travel one Planck length (l_p) in one Planck time (t_p).

Our universe encompasses everything that can be communicated, namely any and all information, emitters, transporters and receivers. The boundaries of our universe are the speed of light and zero time, any events beyond these limits must be regarded as inaccessible from our universe. They are not part of our world.

From this understanding of the Planck time, again, we can deduce that the "flow" of time cannot be smooth, but must be quantum-like. The Planck time is the shortest possible unit of time. It is the minimum time period available for any event and:

> *There can be no smaller increment of progress*
> *into the future than a Planck time.*

If time cannot pass with any smaller increment than a single Planck time then it is impossible for time to "flow" smoothly or continuously. If there are no smaller increments of progress into the future, then after one Planck time, the next amount of progress must be another Planck time. Furthermore, each Planck time must be "separate" from the next Planck time and also be separate from its preceding Planck time, otherwise they would run in to one another and time would flow continuously.

We therefore deduce that, at the Planck scale:

Time must progress with Planck time steps and/or jumps

There must be a temporal "jump", associated with each Planck time, to provide temporal progression.

We must understand that the succession of these time quanta is extremely rapid, almost unimaginably so, and that the universe "stutters" into the future at such a fast rate that we cannot detect the stutter, even with today's best available experimental equipment.

We know that time passes at varying rates, depending on our position within the gravitational fields of massive bodies. We have, therefore, a variable time rate field throughout our universe, since all gravitational effects spread to infinity.

Events unfolding with the passing of time, at any point within this time rate field, could be compared to the TV picture we watch, without noticing the picture changing, with an imperceptible stutter sixty times every second. The reality of a TV picture, is that we are presented with a still "shot" every $1/60^{th}$ of a second, with each "still" being slightly moved on in time from the previous one. From within the viewer's relatively "slow" existence, compared to the rapid screen refreshing, everything in the picture seems to move smoothly, not incrementally – but this continuity is false, it is an illusion. The faster the time quanta (or screen refreshing, as in the case of the TV), the closer to our macro "reality" the illusion gets. But, for the TV, if you view the picture at $1/60^{th}$ second intervals, then you see each individual frozen frame, the reality of the transmitted information.

For our universe, at the Planck scale, where the lower limits of time and space are met, the illusion has also completely disappeared and we would witness the ticking of the universal clock, the reality of our existence!

There are several contenders for the "Theory of Everything" and this idea is new, although somewhat similar to the initial assumptions made in the leading theory of "Loop Quantum Gravity" (LQG).

So, we must diverge from the LQG assumptions and apply these new, philosophical interpretations at (and below) the Planck scale. Space is also considered to be "granular" in LQG, but here, we will see it is only time which is quantised and not length. It is GR which produces this granular effect and only when we have modified it, to reflect only temporal effects, will LQG be able to, realistically, represent space-time at the smallest scales.

There is a philosophical disagreement with time being measured, or compared, against itself. It is often said that you cannot measure time against time, but I disagree. This argument confuses the unit of time with the process of time, the Planck time with the evolution of time, or the energy of time with duration. After all, if time is quantised then it is a series of events one after another and so is entitled to be measured against duration since there are a finite number of Planck times in any duration. This philosophical disagreement is therefore unsound.

Time is the facilitator of change, it allows change, even enforces it. For the process of time to do this, then time itself must also be undergoing constant change. It can never stand still and yet, it seems from our perspective to move constantly into the future, as if it is a smooth, linear process. There is only one way for time to be continuously changing (and for it to move us into the future), it must be wavelike.

A wave is a continuously changing form, yet overall, from a much slower perspective, it looks linear, like an "arrow" moving along its "x" axis. If this wave were fast enough compared to our relatively slow perspective, then it would seem to us to be moving us into the future in a continuous, linear fashion and this illusion has misled us in our attempts to understand time.

Wavelike Time

"If you want to find the secrets of the universe, think in terms of energy, frequency and vibration."

Nikola Tesla

We have until now, simply accepted that *light* is of a wave nature, we have never questioned the absence of an identified cause for this effect. We have just accepted the wave nature of light as a given, but there *must be* a cause for this and we will establish what it is here.

Science has long struggled with the absence of a medium through which the wave form of light propagates. Even today, there is no identified mechanism which produces the wave form of light. We do not know why photons bunch up then spread out then bunch up then spread out, and so on, to produce a cyclical variation in light intensity over each wave cycle and over length or distance. We simply advocate that this is the nature of light, but this thinking (or lack of it), "ducks" the issue and avoids the disturbing consequences that arise from deeper interrogation.

For the transmission of light, in the vacuum of outer space, there is nothing available to produce this wave effect, other than time, itself, so, we have no alternative but to deduce that it is the nature of time that somehow produces the wave nature of light. For the analogy of ripples in a pond, the ripples are the expanding effect and the static medium of the pond is the field through which the ripples travel. The energy of the wave-like ripples disturbs the otherwise smooth medium of the pond. But, the energy of an expanding field of electromagnetic radiation would naturally be smooth and continuous, in the absence of any external cause for the wave effect. So, we must conclude that the time rate field, itself, is wave-like. In this case, the otherwise smooth energy of the light passes through the waves of the medium of time. It is the field of time that is wave-like and any particles passing through this *standing* wave will exhibit a wave nature as a result. Quantum mechanical theory has established mathematically that everything has a wave function, but it has not quite identified the physical/temporal reason why. Our mathematics of classical theory contains many clues as to the wave nature of time. The term "π" crops up everywhere, for no obvious reason. The term gamma "γ" is a "circular" function depicting motion through a sine wave, yet no one seems to have noticed this "circularity" of everything, this wave nature of time.

Time is a *standing* wave, at any fixed position. By a "*standing*" wave, I do not refer to a wave with stationary nodes and antinodes in the physical world, but to a wave in which time itself progresses faster, then slower, then faster, then slower and so on, throughout local space. The "stationary" nodes and antinodes are on the axis of average duration. For any local volume of space, the graph of instantaneous time rate is simply a "straight line" across volume, which goes up and down in a sinusoidal way, remaining straight in the process. Neighbouring points have practically identical time waves except that in gravitational fields the sine wave gradually lengthens with reducing elevation and loses energy (the line goes up and down a bit slower, the nearer the mass). This produces the time curvature within the field, the observed *red shift* and the Newtonian gravitation, as well as the effective space curvature in strong gravitational fields.

Why a *standing* wave in particular? Well, we have already seen that there is a field of time throughout all space and that, locally at least, it runs at the same rate across distance. Even in a gravitational field, the changes in time rate are very small, at least in weak fields. The field is stationary because the time rate in any local region of space (and at any instant) is constant across volume, in the absence of motion.

We seem to have ignored this one, alternative, possibility of a wavelike nature of time which *does* provide a cause and effect, not only for the wave nature of light but also for the time dilation of motion, as we shall see a little later. In addition, it gives us a better insight into wave/particle duality. Whilst it is true that everything has a wave function, this is due to the wave nature of *time!* All radiation consists of discrete "particles", even in transit, and it seems that Isaac Newton's corpuscular theory has been right all along.

Time is a *standing* wave at any "fixed" position. By fixed, I mean any point that does not move through the time rate field with a velocity which is significant compared to the rate of quantum time. In other words, a point which does not move through the time rate field with a speed of a significant fraction of the speed of light … relative to the stationary field.

This idea of a wavelike time is demonstrated by asking the question "Why do light emitters emit light in spherical "shells" of high densities of photons, interspersed with shells of low densities of photons?" As light is generated by a source, say a star, it is created within the local time rate field, so at the scale of electromagnetism, the beam is produced as a cyclical ejection of photons, reflecting the time rates at each part of the local time cycle. There are many more photons emitted at the peaks of the time wave, where time runs fastest, than are emitted at the troughs, where time runs slowest. So, light is emitted as a true, physical wave with ripples of high and low levels of photons ejected into space. This effect is then maintained throughout the beam's progress through space by the *standing* wave of time, which occurs everywhere.

We should note that, during its travels, the journey of any particular photon is instantaneous, since time stops at the speed of light. This is because the speed of light *is* the speed of time and we can now see this very clearly. The photon moves through space at exactly the same rate as the time rate cycles are generated by the *standing* wave. Every photon then "surfs" the time wave at its unique, individual, position on the wave. It is only when the beam impacts our detector that we observe this wave effect with its peaks and troughs of photon density. So, when a light beam passes through any "fixed" position, it appears as a wave form, firstly due to the fluctuating time rate at its original point of photon emission, and secondly, by the wave form of time being in phase with the speed of its progression, throughout the whole of its journey.

Finally, on impact, and as the time rate varies cyclically at our detector, the intensity of light (density of photons), will change with time (our average time) and the detector will show us a wave form of light. For more than one hundred years, we have simply believed our eyes and have dismissed light's wave nature as being "just the way things are", without any further thought.

The standing wave of time is the sinusoidal variation in time rate deduced from the wave form of light.

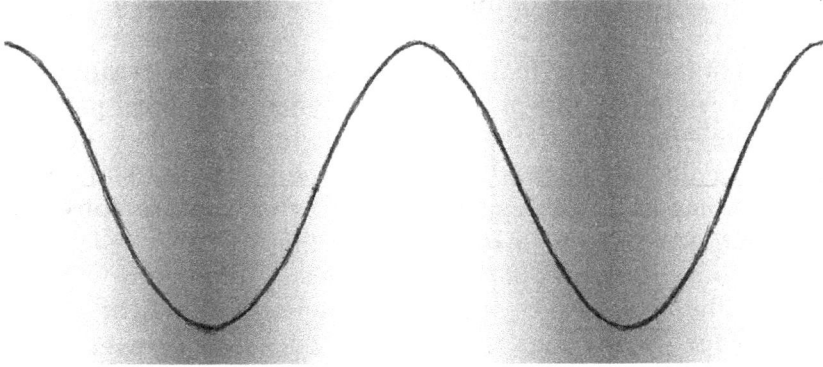

Figure 20: The wave form of light and time

In Figure 20, the light areas have the greatest photon density (light intensity) and the graph is the classic linear representation of this intensity. The light waves travel from say, left to right, and their form is undisturbed by the motion. A photon, at any position on the wave, stays in that position for the whole of its journey. Figure 20 also shows how the wave form of time is, in fact, also the wave form of light that we observe in practice.

Photons are emitted in great quantities at the peak of the time wave, when time is at its fastest rate, compared to the number emitted at the trough of the time wave, when time passes relatively slower. After emission, these bands of high intensity light (high numbers of photons), are maintained by the wave nature of time throughout the void, since the speed of light through space *is* the speed of time, the speed of the wave. From the photon's perspective, it does not experience the wave form of time and it is the wave that drives us all into the future.

The photons travel at the speed of time/speed of light and time has stopped for them, due to inertial time dilation. The variation in Photon density is produced at the point of emission and this perpetual, cyclical effect is then maintained by the stationary cycles of time, throughout three-dimensional space. The photons do not move relative to the time wave, but they are kept in this relative compression/expansion state by the time rate variations over the wave cycles. Of course, this effect only occurs at the speed of light, "*c*". With marginally slower speeds, the photons will "see" *some* of the cycles of time and we will investigate this effect later.

Each photon always maintains the same relative position on the wave, none of the photons experience any time passage during their travels at the speed of light, since they do not experience the wave. It is only we, in our virtually stationary position, compared to the speed of time/light, who are subjected to the time wave and we experience the time.

So, we have the peaks of the light waves simultaneous with the peaks of the time waves and the troughs of the light waves simultaneous with the troughs of the time waves. The frequency/wavelengths of both light waves and time waves are the same and, when we observe the wave form of light, we are actually seeing the wave form of time. We will deal with the different frequencies of electromagnetic radiation (EMR) later in this chapter and you will see there is no conflict between this view of time and the electromagnetic spectrum.

This effect is present at all positions along the path of the light beam since the time rate field exists at all positions throughout three-dimensional space. The light beam is of a wave nature over the whole of its path. We will observe a wave form of light wherever we intercept the beam, but it is ultimately the difference in time rates between the emitter and detector which can determine any *red* or *blue shift* from its frequency at the point of emission. This red or *blue shift* is independent of any gravitational or time dilatational fields the beam may have travelled through, it is only the difference between the time rates, at the start and finish points, which produces any observed red or *blue shift*, (plus any Doppler effects).

A light beam, on its way to our telescope, will enter the gravitational fields of massive bodies and will become *red shifted,* but it will also always exit such gravitational fields and regain the *blue shift* that it had lost. We can see that such occurrences have no effect on the ultimate red or *blue shift* of the light received at our detector, any such shift will simply be the result of the difference in time rates between observer and source.

Time is quantised at the Planck scale and each "moment" is separated from the next by one Planck time. Nevertheless, there must be more to the wave mechanism and we must ask: How can time be quantised at the Planck scale, yet be of a wave nature at the scale of electromagnetic radiation?

If we accept that the Planck time quanta do not, necessarily, occur at uniform, regular "intervals", then we might imagine that they bunch up, then spread out, then bunch up, then spread out and so on, in a cyclical way relative to the average "spacing" of previous and future time quanta. The "arrow of time" grows faster, then slower, then faster, then slower, and so on, firstly as a step function at the Planck scale, but also as a wave function at the scale of EMR. This is the only possible way in which time can be both quantised and wavelike, and it does have to be both.

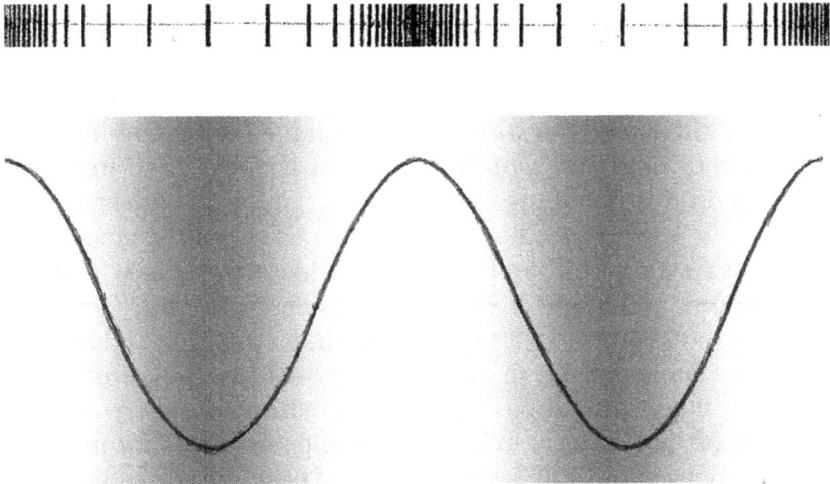

Figure 21: Quantised and wavelike time

Figure 21 demonstrates pictorially, how this works. Of course, I have reduced the number of time quanta, for the purposes of clarity. There are many millions, even billions of quanta occurring over each EMR cycle. Also, there is a limit to the degree of "compression/expansion" of the time quanta spacing and, although the diagram demonstrates the principle, it needs further, detailed explanation. This will be given later in this chapter. From Figure 21, we can see how the cyclical variation in quanta "frequency", at the Planck scale, produces the sinusoidal wave form of time and light, at the larger, classical scale. This cyclical variation in quantum "spacing", at the scale of electromagnetic radiation, shows how the wave form of time is generated. There is no conflict between the quantum nature of time, at the Planck scale, and the wave nature of time, at the macro scale. We will deal with the effects of the EMR spectrum very soon.

In the absence of any other driver for the wave nature of light, we are forced into this conclusion, since there is no alternative explanation and no other available cause for this effect. So, this is the way it *must be*, unless anyone can come up with a better idea? The variable temporal spacing of the time quanta is the driving force of our universe and, although the cause of the quanta cannot be explained with our present knowledge, this should not stop us from making these deductions. here. The explanation for why the quanta behave in this way is given later, in chapter 12. In doing so, we are simply removing our ignorance one step at a time, so we must "park" this philosophical issue and deal with it at some later date, when we have further understood the nature of time and *why* it does what it does. If we insist on understanding everything, all at once, we will never make progress. If I were to hazard a guess, I would suspect the sinusoidal spacing of a binary, quantum-like time is inevitable in any vacuum, although this would result from only a slight possibility, but that this possibility is perhaps the only stable, long lasting one amongst the many other possibilities, however more likely they might be. All possibilities happen eventually since the definition of "possible" is to have a finite probability, but perhaps only our "possibility" has lasted (so far). These ideas are pursued in more detail later.

Before we proceed you may question...

> "Why should we accept the wave nature of time, rather than the wave nature of light? Are we not simply transferring our ignorance, our blind acceptance from one aspect to another?"

Well, I think not.

Light is a phenomenon within our universe, so its wave form needs an explanation. It must have a knowable, physical cause and effect. Time is more fundamental. Time *may* have a knowable cause and effect, but we may never achieve a complete understanding of it. It is the fundamental part of our universe, the driving "force" of our world, but light is a phenomenon *within* it. I suggest that because of this, it is far more acceptable to resign ourselves to the wave nature of time (for the time being), rather than the wave nature of light and that we are right to ask the question, "What is the cause?" In this way we can make progress.

Most importantly, this view of the nature of time gives a direct causality for the wave nature of light which, in the absence of any other possible cause, is itself compelling evidence that this understanding is correct. The next section will add even more to this weight of evidence, by showing that the causality for inertial time dilation is also implicit in this view of the nature of time.

Causality of Inertial Time Dilation

Mainstream science cannot explain how or why light has a wave form, nor can it explain how, or why, time slows down with relative motion. It proposes no cause and effect for either, it merely proves their inevitability and accepts the facts, without question.

There is considerable confusion over time dilation and whilst the mainstream does seem to accept the reality of it, this begrudging acknowledgement has been "fudged" by the unnecessary, additional concept of length contraction. Confusion is exacerbated by the irrational belief that all motion is purely relative and, in this situation, it seems, no one has felt it necessary to investigate causality.

I will do this now.

I have just explained in principle, the cause and effect for the wave form of light, so now let us see how this same proposal, for the wave nature of time, might also cause time dilation in accordance with Lorentz. We must ask:

"What is the physical cause and effect of inertial time dilation?"

Special Relativity is based upon one premise, that the speed of light is invariant. Light speed is the same when observed from any frame of reference. This fundamental rule when applied to the geometry of motion, inevitably means that time dilation will occur, as a result of the motion, in accordance with the Lorentz transformation. Although the geometric proof of time dilation is based on this invariance, it does not explain the mechanism involved, the cause and effect. We have, so far, simply accepted the inevitable result from holding the speed of light to be constant in all 'frames, but we have not asked *why* or *how* this should be the case. In establishing the cause and effect for the time dilation of motion, we are also establishing the cause and effect for the constancy of the speed of light, since these phenomena are one and the same. If you have the one, then the other is inevitable.

In the preceding explanation for the nature of time, we deduce that time passes in quantum increments, rather than in a smooth, continuous way. As you approach the speed of light, the time taken to traverse each Planck length reduces to one Planck time. Remember, it is time that is quantised, not length. The Planck length can be subdivided as many times as there are Planck times. For sub-light speeds, there are many Planck times required to traverse just one Planck length. The Planck length is simply the *maximum* distance you can travel in one Planck time and this maximum is a result of the reduction of the number of time quanta (Planck times) to *one*, in accordance with Lorentz. At the speed of light, this single Planck time is your surf board as you surf space on the wave of time. You never jump to a second one, you never move on the wave, so time stops for you.

At low speeds, there are many time quanta happening compared to your slow motion through the field of time, but, as you accelerate to very high velocities of a significant fraction of the speed of light, the rate of your physical motion becomes significant relative to the spacing of the time quanta, the rate of time passage. Ultimately, at light speed, we might think we get from one Planck length to the next, then leave it, *before* the next time quantum happens, *before* the (local) universe takes its next quantum jump into the future! This is perhaps one way that a quantum-like time might slow down with speed, so it could be the basic mechanism of time dilation, but we will soon see that things are a little different than this.

If motion continues at light speed, then you will never jump with any time quantum during your travels, since you will continue to miss all the jumps into the future. At light speed, time has stopped for you and your speed has therefore become infinite, relative to the "stationary" frame. You will get from A to B instantaneously, whatever the distance involved, therefore, the distance (length) has become irrelevant to you.

The traditional way of saying this, is that "Lengths have contracted to zero-length, in the direction of motion", but this is clearly ridiculous in the real, physical world. Even though this "counter-intuitive" (ridiculous) view gives us sensible results, we must realise that to achieve the progress with our understanding, then we must dispose of our belief in length contraction. Until we do so, we will remain at an impasse. Lorentz length contraction is merely a useful, geometric idea which can be used to give us the right answers, but it is not a real phenomenon. The only real phenomenon is the time dilation which, on its own, gives us the same correct answers.

We know that time slows for the moving (accelerated) entity and we can now see more clearly why length has become irrelevant at velocity "c". At light speed, we are "skipping" all future time quanta. We are missing all the quantum jumps into the future, as we move at the speed of light and, wherever we stop, no time has passed for the traveller. This is best understood from viewing our distance travelled, "S" as:

S = The number of Planck lengths covered × Planck length l_p

But also, and only at the speed of light:

S = The number of time quanta (Planck time) skipped × Planck length, l_p

These two statements are equivalent, since it takes one Planck time to cover one Planck length at the speed of light and:

*At the speed of light, the number of Planck
lengths travelled equals the number of Planck
time units passed for the static universe, but
which have all been skipped by the traveller.*

The further, distance wise, we travel at light speed, the further
into the future the universe has travelled, by the time we stop.
The "static" universe has "used up", or "gone through", all the time
quanta that the traveller has skipped and, as a consequence, it is
that much older than him, when he slows down. In order to allow
the skipping of the time quanta due to motion, the time quanta
that you miss must somehow be unavailable to you, when they
occur. We will see how this happens in just a moment.

The skipping effect has to occur for any and all directions of
travel, not just in one specific direction. Inertial time dilation
occurs irrespective of the *direction* of motion. Only with a
spherical nature of space-time, can you have inertial time dilation
independent of direction. Clearly, there is no real, physical,
spherical nature to the void, since it is continuous, but it is simply
that the behaviour of time, as you travel through it, presents us
with this spherical nature. Space-time is effectively a spherical
matrix of quantum time jumps from any particular perspective. In
practice, this means that whatever point you are moving away
from, at any moment, can always be regarded as being at the
centre of a sphere and this is what is meant by a spherical nature.

Since the quantum time jumps are extremely "rapid", it is not
surprising that the time dilation effect, due to motion, is very small
for slow speeds through the time rate field. Slow velocities are
negligible compared to the rate of the quantum jumps and the
traveller experiences practically all the time quanta. At the other
end of the scale, it is obvious that as our speed approaches "c",
then speed is becoming very significant, relative to the rate of the
quantum jumps. This is exactly what the Lorentz transformation
tells us. Lorentz shows that time dilation is very small for low
speeds, but increases until it eventually stops time, altogether, at
the speed of light. This happens in a circular way, as a circular
function and it is movement through the sine wave of time that
gives us this "circularity".

We must understand that the limit of the speed of light is actually imposed by the finite rate of time. There is a direct causality between the two and both the speed of light "c" and universal time are finite, interrelated values. Their maxima and minima are opposite and coincident.

> *But how does the skipping of time quanta happen? What is the mechanism?*

Well, since the time quanta are, by definition, separate, then it follows there are temporal gaps between them. The section on *The Binary Universe* establishes the details of this arrangement. For now, we will limit ourselves to analysing the effects of moving through this temporal, quantum field.

We must not make the mistake of trying to visualise physical movement across these quanta, since it is the quanta themselves which produce our quantum jumps into the future. They allow, even drive, all events including motion. They are not just an effect of time, they are time itself. So, after each jump, we find ourselves suddenly at a future physical location, as if deposited at a point ahead in our direction of travel. For the moment, we will ignore what happens between jumps, as it will be dealt with, later. Now, it cannot be the case that the time quanta are like individual "flashes" (jumps), spaced apart physically by one Planck length, like some sort of physical space-time "grid". Space-time is *not* "granular". The jumps have to occur everywhere to enable stationary, or slow moving, entities to experience all (or most) of the time flashes. The jumps must permeate the whole volume of space, at all scales, and the complete volume of (local) space jumps into the future, in unison.

So, we now have a problem. For our previous idea of skipping, we cannot get to a physical position, then leave it before a jump occurs, yet we have to explain how we can miss, or skip, any time quanta (flash) with linear motion. We must consider both aspects of motion, firstly a wave approach and, secondly, a quantum approach.

Travelling through the wave

If we are "stationary", the wave form of time moves us into the future in a sinusoidal way, faster, then slower, then faster, then slower, but this is *all* that is happening. We just sit there and move through time. As we start to move physically, from one position to another, it takes time to do so and so we cannot get to any position at the same time, at the same point on the time wave, as we would have been on, had we stayed where we were.

There must be some "loss" from the action of moving, a price to pay for the kinetic energy of motion and this energy must come from somewhere. Remember, the maintenance of constant linear motion does not require energy input, but the motion, the changing of position, necessarily will require energy, if time is quantised. If we are stationary, with the time wave moving us into the future, then that is all it has to do, but as we move through space, it is no longer possible for us to experience the precise time wave as a stationary entity and we lose some of the time in our moving. Our movement has literally "taken some time" and we have traded a small amount of temporal progress for physical progress (speed). Clearly, this effect is dependent upon speed, not position or acceleration but simply, speed. Since the time wave is very fast (rapid cycles) and our motion is, relatively, very slow, then this effect is very small for slow speeds, but as we approach light speed, we are now trading a significant amount of time for increased velocity. Finally, at "c", we are "taking all the time" to move us through space and we no longer move through time itself. Now we will see why.

There are two distinct effects for a moving entity. The first one we know that the motion results in some distance being covered, in a certain time, depending on the speed, but there is a second effect from this same motion through the wave form of time.

We must understand that the time rate, although varying sinusoidally, is constant across distance, at any instant, throughout any local volume of space. We all move into the future at the same rate wherever we are in the room, so the time rate is the same across any local distance. The time wave can be represented by a straight, horizontal line on a graph of time rate (which goes up and down across distance). This *standing* wave of time has a horizontal axis of "average" time.

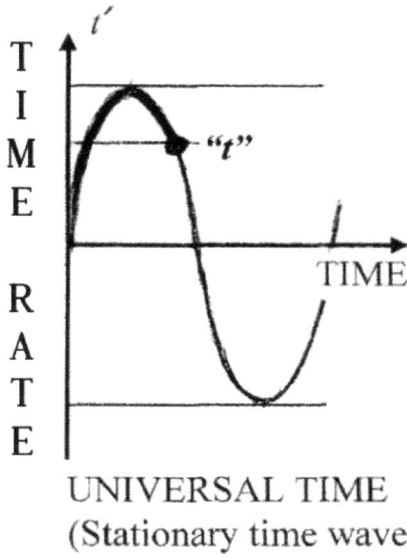

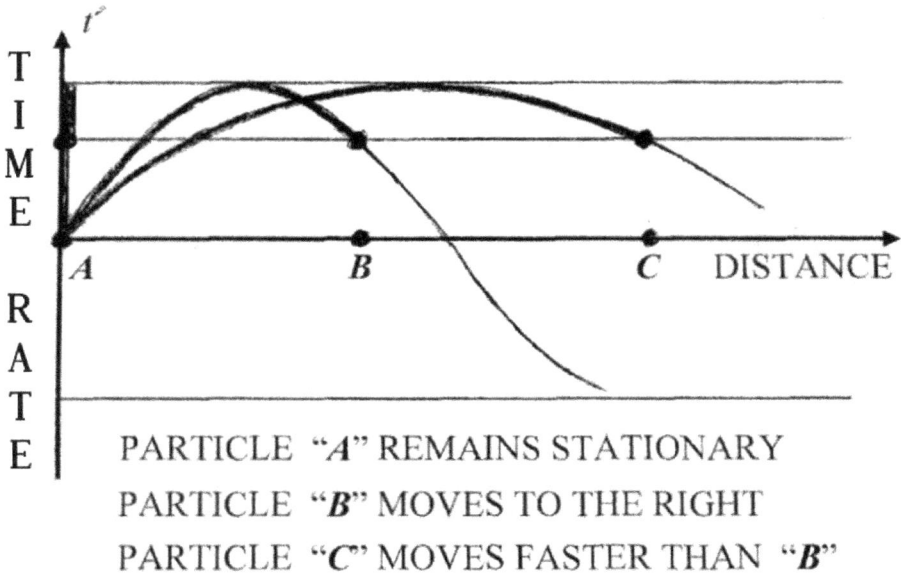

Figure 22: The Cause of Inertial Time Dilation

Figure 22 shows three particles, A, B and C moving at different speeds through the stationary, oscillating field of time and is a snapshot of the situation for all three particles after time "t" (Static time "t"). The *standing* wave of universal time is shown at the top of the figure. This stationary wave form of time has progressed from the origin, up to the upper limit of the time rate and then down to position "t" on the sine wave at the top of the diagram. This duration contains a fixed number of time quanta.

We can see that stationary particle *A* has simply experienced the universal time wave, but particle *B*, which is moving, has reached a physical position further to the right by the time the horizontal line has reached time *"t"* on the stationary time wave.

Clearly, the wave form of time, experienced by particle *B,* has been stretched by the motion, the particle has experienced a longer wavelength of time, a slower time rate as a result. Again, this effect is more pronounced for the faster particle *C* and the time wave has elongated even more. The point is that the number of Planck times that have occurred for all three particles, up to this moment, is the same. However, this fixed number of Planck times has to be spread over each wave, however much it has been stretched to different degrees, due to their different motions. This is the mechanism that *red shifts* time with increasing speed. Time, as experienced by a moving entity, is *red shifted* (Transverse *red shift*) by the motion, this effect will be more pronounced for greater speeds. The relationship between this *red shifting* and increasing linear speed is clearly a circular one, due to the wave nature of time and, by inspection, will result in Lorentz time dilation, which is of course, the equation of a circle.

You may think that, because each particle has experienced the same number of Planck times, they must have moved into the future by the same amount. But it is the wave that drives us into the future, not the time quanta. The time quanta are merely opportunities for temporal change which may be taken up, or not. They are packets of energy, which can either be used as temporal energy or kinetic energy and the moving particles have used some of them for motion.

> *A moving entity still experiences all time quanta, but it uses some for speed instead of for temporal progression.*

For clarity, I have exaggerated the effect by showing the movement over a significant part of a single wave, but the effect is only this significant for very high speeds close to light-speed. For slow speeds, compared to the speed of light/speed of time, there are many cycles of time happening over distance covered and this *red shift* effect will be very small indeed.

As speed increases, the *red shifting* increases and, at the speed of light/speed of time (the speed at which the horizontal line goes up and down), the wave form is completely red shift*ed*. The time wave becomes a straight line and no wave effect remains for the moving entity. Time no longer passes.

It is now clear why the degree of this *red shift*, with increasing speed, is described by the Lorentz transformation, because the sine wave is a circular function.

Figure 22 shows the physical connection between time and space. We can actually "see" the link between them. We can grasp it, intuitively, and we do not need any math, or geometry, to understand the effect, just our mind's eye. This has become possible from our new understanding of time as a stationary wave. The mechanism, the causality, is thus proven by inspection.

The transverse *red shift* of time increases with increasing speed and for faster entities, the time wave is "smoother". It has a less pronounced rate of change of time rate, less temporal "jerk". The curve is never as steep for the faster entity, at any point on the wave. Importantly, we can now also "see" that, at the speed of light, the *static* wave form, from the moving perspective, is completely compressed and *blue shifted*. We can be comfortable with this idea, because we now know there *is* a preferred reference frame, the *standing* wave of time.

From Figure 23 we can see how the transverse *red shifting* of time increases with increasing speed from zero to "*c*". For a stationary entity, the wave form is simply the oscillation of temporal progress up and down the vertical axis, just like particle "A". For the moving particles, the vertical axis is time rate and the horizontal axis is distance travelled. For the graph of universal time in Figure 22, the vertical axis is also time rate, but the horizontal axis is duration.

For a stationary entity:

For a slow moving entity:

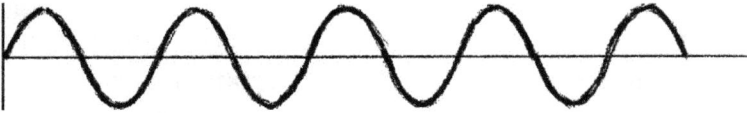

For a faster moving entity:

For a near light speed entity:

At the speed of light/time:

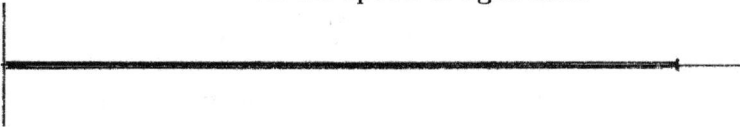

Figure 23: Increasing *red shift* with increasing speed

Travelling through the quanta

The last section shows how time itself becomes *red shifted* by motion, but it is the quantum nature of time that means it loses temporal energy for the traveller. It is critical to understand that, with "zero" velocity, we use *all* of the Planck time quanta to move through time but, by the time we attain light speed, we use *all* of them to move through space. For speeds between zero and "*c*", we experience temporal effects from *some* of them, depending on our speed, in accordance with Lorentz. So again, we can deduce that we are simply "trading off" temporal progress for physical progress, between the speed limits of zero and "*c*". We are trading off time for speed.

Again, if we are stationary, we just sit there and move through time, so all the time quanta are "used" to move us into the future at the finite rate of 1.855×10^{43} Planck time units per second of universal time. As we start to move, most of the quanta are still used to move us through time, but some are now used to move us through space. Ultimately, at the speed of light, all of the quanta are used to move us through space, so none are available to move us through time.

As we start to move through space, the quanta not only have to deal with propelling us into the future, but now, some are needed to propel us to a different location. There must be a price to pay for this. You cannot get something extra for nothing extra. Remember, we are not inputting energy to maintain constant motion, but the effect of regularly appearing in a different position will have to be paid for by the quantum effect of time, itself. There is nowhere else for the energy to come from. The kinetic energy of constant motion is topped up, continuously, by the perpetual use of the Planck time quanta, so those doing the topping up are unavailable to use for temporal progress. They can only be used once.

With a quantum nature of time, as indeed with any quantum-like phenomenon, what happens, for any change in overall effect, is a loss of quantum activity. Some of the quanta get eliminated, in terms of their effect. A quantum either has an effect, or it does not and there is no question of degree. A quantum effect is a binary effect, a "0" or a "1", an "On" or an "Off". The effect is absolute, so the only way it can change an overall effect, is for some of the quanta to be eliminated from that effect and not for any individual quantum to vary in any way.

What is happening, as we accelerate from zero speed, is that some of the quantum energy, in the form of time quanta, is needed to produce a change in position, so those are rendered unavailable to change our temporal position, to move us into the future. Ultimately, we travel so fast, we need all the quantum energy to keep changing our physical position, so none of the quanta are left to move us through time. Like the photon, we stay on the same point on the time wave, on the same quantum and we do not jump to the next.

We can deduce from this (if we are right) that temporal energy is equivalent to kinetic energy and that no Planck time quantum can be used for both, but only for one or the other.

For constant, linear motion at light speed then:

> *A quantum change in position is the energy equivalent of a quantum movement into the future.*

So, the energy of one Planck time, applied to a stationary entity to move it into the future, is equivalent to the energy required to move that same entity, at light speed, by a distance of one Planck length.

The work done in an event is the change in energy over the time of the event, so the work done in one Planck time is the energy expended for that time quantum. For either temporal or physical progression we can say that the work done in moving a stationary entity through one Planck time is equivalent to the work done in moving that same entity through one Planck length at the speed of light.

Now,

Work done = Force x Distance moved,

But also,

Force = Mass x Acceleration,

and so,

Work Done = Mass x Acceleration x Distance moved,

So, for a light speed Planck mass,

$$WD_p = m_p \times a \times l_p$$

For a single quantum time progression, acceleration is:

$$a = \frac{l_p}{t_p^2}$$

And so,

$$WD_p = m_p \times \frac{l_p}{t_p^2} \times l_p = m_p \frac{l_p^2}{t_p^2} = m_p c^2$$

Since this is the quantum work done in one quantum Planck time, we can also say:

$$\text{Planck Energy, } E_p = m_p c^2$$

and therefore, for any mass, at all scales,

$$\underline{E = mc^2}$$

So, from this new understanding, we can see that:

"E" is the energy required to move a static mass "m" through time!

This is also called the "static mass energy", it is equal to twice the kinetic energy of the mass at the speed of light. (We shall see why it is twice, in a moment).

You may ask,

> *"What has a static mass energy to do with the speed of light?" "The mass is static!"*

Well, this is why - the speed of light comes into it, because it reflects the limit to the number of available Planck times, the finite rate of Planck times. The speed of light is a measure of the speed of time.

We have just derived this well-known equation independently, from our new understanding of the nature of time at the quantum scale. So, this successful derivation, together with the fact that we *know* time slows down with speed, *proves* that the concept of trading time for kinetic energy is a valid one and therefore that:

Time is energy!

So, we *are* right and progression through space is, indeed, the energy equivalent of progression into the future. The function of the time quanta is simply transferred from one duty to the other, in varying amounts for different linear velocities.

We can now say that:

Time energy = mass energy

or,

$$E_{TIME} = E_{MASS} = mc^2$$

A faster moving entity will simply use a lesser number of time quanta to move into the future, but a greater number of time quanta to move physically through space.

There is a finite number of quanta available. The combined number of quanta used, for both effects, is a constant, it is invariant, since the rate of production of time quanta is constant at 1.855×10^{43} per second duration of universal time.

This is the fundamental reason why light speed is the limit of velocity in our universe.

The limit of velocity "c" is the result from the finite rate of time, the finite number of Planck times (over duration).

At the speed of light we are using up all the time quanta at the rate they occur, to move us through space, so none are left and we can go no faster.

We can also say:

> *The energy of inertial time dilation is equal to the number of Planck times used for speed (kinetic energy), instead of for temporal progression.*

Furthermore, we know that,

$$Mass = Energy$$

and we now know that,

$$Time = Energy$$

therefore, we deduce that,

$$Mass = Time$$

Mass itself, or matter, must be made up of energy from the field of time, which is why the field reduces in energy, in the presence of matter.

This confirms our understanding of the causality of mass induced time dilation in the earlier chapters of this book. Some of the time quanta in the field are used for the internal kinetic energy of the particles within the mass, so the field strength reduces and time slows down local to matter.

Any individual time quantum can be used for either moving us, physically, through space or temporally, through time, but not for both. So how is any particular quantum "selected" to move us through space, as opposed to through time?

The sheer number of time quanta occurring (10^{43}) over "sensible" movement only requires a random selection of quanta to even out the "spread" over the total number of quanta occurring. It is the wave form of time, which then produces the overall effect of the Lorentz transformation from this statistically "even" selection of quanta.

So,

> *God does "play dice", but the odds are stacked well in his favour.*

Now, we have found that time is a form of energy and speed is also a form of energy, kinetic energy so, if we accept that the available energy is not limitless and is finite, then it must be the demand from our energy of motion that somehow ensures a certain number of quanta are used to move us through space as opposed to through time. Let's take a single Planck mass moving at some velocity v. The Kinetic Energy of motion is:

$$E_{KE} = \frac{m_p v^2}{2}$$

and at the speed of light:

$$E_{KE} = \frac{m_p c^2}{2}$$

Now for a velocity of "c", this is only one half of the Planck energy due to time (mass energy) and, similarly, for any speed "v" the kinetic energy will be only one half the energy of the time quanta used for speed.

So, the next question is:

> *"Where has the other half of the energy gone?"*

Since *all* the time quanta (energy) are being used for kinetic energy by the time we attain light speed, we should expect the kinetic energy at "*c*" and the static time energy to be the same, but they are not. The static time energy is *twice* the kinetic energy, at the speed of light.

The next section will show that kinetic energy is always one half of the time energy, because we live in a binary universe and the junction of the two parts lies in the phenomenon of time. Events, although equal and opposite, occur *separately* in each binary part and the total kinetic energy is shared equally between the two parts. It cannot cross from one part to the other. Time on the other hand is *directly applicable* to *both* binary parts and both parts move into the future at the same rate, as they must. We will discover why this is very soon.

The Binary Universe

At the Planck scale, it is not that space itself has a limit below which it cannot exist, but it is simply that the temporal "spacing" of the time quanta does not allow transit across anything more than one Planck length, per quantum jump at light speed. That is to say, if you travel at the speed of light for one single time quantum then you must, of necessity have travelled one Planck length. That is the definition of these Planck units.

The void is still available *below* the Planck length, since we can travel very small fractions of it, for each time quantum at low speeds. Again, the Planck length is *not* a fundamental unit. The Planck *time* is *the* fundamental unit, the *"Fundamental Quantum"* and everything else is secondary, a resultant or emergent.

In attempts to develop a theory of quantum gravity, GR has been combined with QM to produce "Loop Quantum Gravity". The mathematics of LQG interprets space-time as increasingly "foam like" as we get down towards the Planck length, but this is because GR is a geometric interpretation of space-time and the math cannot cope with the reality of sinusoidal time changing to quantum time as we focus down to the Planck scale. It is impertinent to believe that we can use a classical, macro theory of behaviour to describe events toward the Planck scale, where the classical, macro rules of behaviour no longer apply. Contrary to the approach used in LQG, the new theory does not focus on the dimensions of physical space, but suggests we look to the nature of time at the temporally "small" (fast) scale to explain events in the quantum world. Just as with length contraction in Special Relativity and with the curvature of space in General Relativity, the B.U.T. proposes that only time varies with position and by each moment, even at the quantum scale, and that the geometry of space is irrelevant. Time is a real phenomenon, but space is not, it is ultimately nothing, a resultant from time passing.

To digress, I may as well make this point here: It is not possible to travel faster than light, or more fundamentally, faster than the speed at which the clock stops. So, we must be consistent with this, when it comes to black holes and Planck times. If the clock has stopped on the event horizon of a black hole and events have ceased to occur relative to us, then why do we try to imagine movement through the event horizon and below it, when the clock has stopped there? If time has stopped at the event horizon, then there can be no further movement inwards towards the centre, as far as we are concerned, therefore there can never be any singularity, relative to us. The conventional view of the entity moving through the event horizon is flawed, because it views time as purely relative. It ignores the fact that time dilation is absolute. When time stops, it really does stop, in absolute terms, and a zero time rate is zero from any *and all* perspectives. This statement is equivalent to our notion of light speed invariance, viewed from any *and all* reference frames. It is the same notion.

There can never be any progress beyond the event horizon. This is not just a relative effect from our external perspective, but it is also real in the moving frame. It is no good arguing that anyone falling in still experiences time normally, of course he does, it is simply that he is unaware that his clock has stopped and that the rest of the universe is progressing into the future infinitely faster than him.

If we place ourselves in the frame of reference of the entity, which falls through the event horizon, there simply cannot be any progress toward the "singularity". Time really has stopped, relative to the rest of the universe, so any progress will take until the end of the universe and longer. By the time any movement inwards happens, the black hole will have evaporated with

Hawking radiation. A person falling in will not realise, before this happens, as he will have used only a handful of Planck times, by then, to move him into the future. He has been starved of practically all the time quanta which have been used to spin the black hole's huge mass around at enormous speeds.

It might be argued that, since both the traveller and his immediate environment have the same frame of reference, then motion can continue relatively between them, but this still ignores the fact that the frame of reference, outside the black hole, has an *infinitely* faster clock and events outside will overtake *any* movement towards the "singularity". It is correct that he still experiences time normally, but the end of the universe will occur before he makes much progress. For this reason, not *one* of all the black holes in the universe, for however long they have existed, has a singularity inside it (yet). The mathematical impossibility of such entities is thus avoided.

From this new understanding we conclude:

> *There is no such thing as a black hole singularity*

The mathematics which predicts a singularity is perfect in geometric terms, but it is naive in temporal terms. From the original formation of a black hole, the point at which the event horizon forms, any additional matter and energy must be trapped, forever, between the initial Schwarzschild radius of the black hole and the final, larger radius (created by the extra matter having fallen in, later on). We do not even have to consider the physical stability of this "shell", since it is stable, by virtue of its eternal nature. This stability is analogous to the eternal perfection of a light image travelling many light years, because it experiences no time progression. From our frame of reference, we will wait for eternity for a singularity to form in a black hole, *so it never will*. Effectively, the volume inside the event horizon is isolated from our universe. It is no longer part of the world that created it.

Physicists have an aversion to the idea that "information" can be "lost", but there is no compelling reason, no natural law that means this is impossible. In any case, there is still the time dilation field around the black hole which describes, or at least quantifies, all the information contained within it. You can establish all the fundamentals of the matter/energy, within the black hole, from this time dilation field.

Allegedly, the universe began with the Big Bang and all "information" came from nothing. It may be that all information will, eventually, be lost when (and if) the universe comes to an end. There is no reason to suppose this must happen all at once, no reason to reject the possibility that some of it may be lost earlier than the end of time, in small, discrete quantities. In fact, it could be argued that the event horizon *is* the end of time, locally at least, and there is no turning back.

Current thinking envisions that all the information, for the matter falling into the black hole, is defined by the entropy of the surface area of the event horizon and, of course, this makes perfect sense for matter/energy, which never progresses beyond the event horizon. The additional information has had to "pile up" at the event horizon, queuing to go further in, whilst the outside universe takes its course.

Black holes apart, it is also nonsensical to apply General Relativity near the Planck length, where the curvature of time has been fragmented by the increasing significance of the time quanta. Time curvature is obliterated before we reach the Planck scale and this is what gives us our mathematical quantum "foam". We are being carried away by the mathematics and losing track of deductive reasoning and common sense philosophy. Nowadays, common sense is often ridiculed and those who trust it are sometimes criticised for not accepting the "counter-intuitive" aspects of science. I believe, deeply, that if we call anything counter-intuitive, it is an admission that we do not know enough about it to understand it, not that it defies understanding. *Nothing* defies our understanding and, potentially, we can expand our knowledge to understand everything.

Despite the fashionable (and almost total reliance) on mathematics, common sense must always prevail, although we may need to expand our knowledge to apply it to new ideas. Certainly, mathematics will always prevail, but only as, and when, it is appropriate to apply it.

GR is based upon the sinusoidal nature of light/time, so it is entirely applicable at the macro scale, but its application gives an increasingly inadequate description of space-time at the smallest/fastest scales, below the scale of the wave form. Inherent within GR are the classical laws of nature and these are irrelevant towards the Planck scale. Chaos (foam) reigns, if we try to use the rules of classical physics, which include sinusoidal light/time and the curvature of space and time.

The reality below the Planck time is a "non-reality", since only time progression with Planck times exists within our universe. Space-time, below this scale, is outside our universe, so it is incapable of description by the laws of our universe. To attempt to mathematically analyse anything below this scale is like trying to analyse events below the event horizon of a black hole (and is just as futile).

The nature of time changes from a perfect, sinusoidal time rate in flat space at the macro scale and is subjected to an increasingly quantum-like "interference" with the sine wave as we view smaller scales, until we ultimately get down to the single Planck time. Clearly, there is absolutely nothing sinusoidal about this fundamental unit. It cannot have any curvature on its own. Curvature exists only at the larger/slower scale, where there are enough quanta to create the wave which can vary in infrequency and present curvature, to allow gravity to act. How many time quanta are necessary, to provide us with a quantifiable degree of time curvature, I will leave to the mathematicians but, clearly, there needs to be a certain (large) number before we can sensibly use GR to analyse it. Mathematically, we may be able to model behaviour right down to the Planck time, but it will not be entirely by the application of GR. Quantum theory will be required to gradually and *completely* "take over" from GR, as we get down towards the fundamental unit of time. Below this fundamental building block, there can be nothing in our universe.

Well, we have just destroyed half the universe! So, now we had better create. But what is the way forward? How do we explain the "gaps" between our Planck times? How do we "fill" them, or give meaning to them? There are questions we must ask about the detailed, micro chronology of the quantum time jumps. We have established that time is quantised and that there must be separation between every Planck time, to avoid each one running into the next, producing a continuous time flow. If you think about it, this separation can have only one possibility, that *somehow* there must also be temporal progress *between* Planck times, within the separations, otherwise the universe would be required to "re start" at each new Planck time and there is no causality for that.

Assuming the gaps between our Planck times have the same laws of nature, then the only way we can justify this temporal separation is with another Planck time, but this extra time quantum is outside our universe! Nevertheless, this is the only arrangement that gives us a workable space-time continuum with quantised time.

Considering our time quanta and the temporal "spacing" of them relative to each other, we must understand that events are presented to our universe in quick "snapshot" periods (Planck times), but with nothing in between them. But now we have established that time also progresses *between* our Planck times. Linear motion for instance would appear to progress with a "stutter", to anyone who had a strong enough magnifying glass and a quick enough eye, but that equipment and that eye will never exist. Our eyes and our equipment exist *within* the sinusoidal time rate field and we will never be capable of "seeing" below this scale or faster than this. We might use experimentation to detect effects predicted by this view, but here we will *deduce* what must be and "see" with our mind's eye. In fact, the mathematics of quantum mechanics has already broken much ground on this trail and many of the interactions in QM can be explained by the temporal aspects of the B.U.T.

Within the quantum gaps, constant linear motion *must, somehow,* continue in the same direction across the physical void to get to the next position by our following time quanta, in order that we are presented with the next "flash" or "jump" moved onwards in time. But this happens *outside* our universe and it must be repeated for every single Planck time, *everywhere!* We must conclude there is another, identical but "negative" part to our universe existing between our Planck times.

Our universe is a binary universe!

A quantised or digital, binary universe!

Planck time-jumps, from universe "A", produce progressions of time in universe "B" and Planck time-jumps, from universe "B", produce progressions of time in universe "A", ad infinitum. Time in each binary part then "mirrors" events in the other part. So, a Planck time quantum in "A", must move things on for "B" via jumps from A to B and vice versa.

But this is not the whole story. We know there must be temporal progression, within the durations of Planck times, in both "A" and "B". Time does indeed pass during Planck times to allow progression into the future, but now we see that this progress is only released to universe "B" at the jump at the end of the Planck time in "A". A Planck time is not just a frozen flash within its half of the universe. Events, built up during its duration, are valid within its own binary part, but are then released to the other part, at the end of the next jump. It must be this way for time to progress continuously (as it must) for the binary pair, yet in a quantum way for each binary half. The Planck times have to possess some "duration", since they cannot have zero duration. They would not happen at all, if that were the case. It is the combination of Planck times and jumps, in the binary pair, that creates the progression of time, for the whole. These jumps in time occur at the interchanges between "A" and "B". They release the events built up during their preceding Planck times. These releases happen in the opposite binary part.

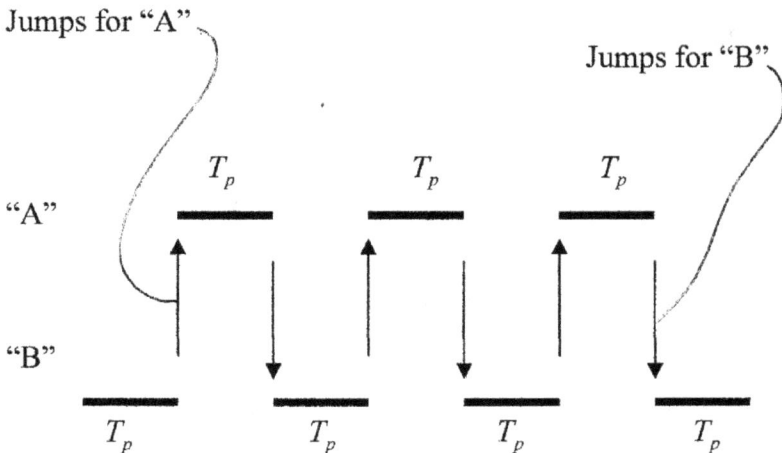

Figure 24: Quantum time progression

Running through the diagram in Figure 24 in chronological order, step by step, from left to right, we have:

- Planck time in "B" allows events in "B.
- Jump from "B" to "A" releases these events in "A" and "A" jumps ahead in time
- Planck time in "A" allows events in "A".
- Jump from "A" to "B" releases these events in "B" and "B" jumps ahead in time.
- REPEAT.

But this "God's eye view" is from "outside" of the binary parts, so what do we experience within our binary part, say from "A"? Again, step by step we have:

- Jump from "B" received by "A" releases events in "A" and "A" jumps ahead in time. *The jump*
- Planck time in "A" allows events in "A". *The run*
- Jump from "A" to "B" (No effect in "A").
- Temporal gap in "A" while "B" progresses. *The pause*
- REPEAT.

We have a repeating *"jump/run/pause"* temporal sequence in each binary part of the universe. The pauses are "opposite" the runs in the other part and the jumps are instantaneous. They have no effect on the binary part from which they originate. The concept of instantaneous is quite valid here, since we are considering time itself and there is no resistance, inertia, or any other delaying effect, when it comes to the phenomenon of time. Some might argue against the idea that we can have periods, or pauses, when time does not progress, but we must remember, these quanta are "energy" quanta at this scale, it is only at the larger scale of the wave that the process of time evolves and comes into being. This is on such a tiny/fast scale that we can never notice the universe "crackling" or "sparkling" into the future, at the Planck scale. In any case, we get the time progression, from our opposite binary part, before our next time-quantum occurs, so all is well at our large, slow scale of existence.

Events, in our part of the binary pair, exclude the progress made by the adjacent Planck time in the other binary part. We get all the benefit of the events in the other Planck time from the jump, but we do not experience the event. This is why the kinetic energy in each binary part is only one half of the time energy for the binary pair. The Kinetic Energy in either part is due solely to the progression of events (motion), during that part's Planck times, and is, therefore, only one half of the total kinetic energy in the binary pair. Overall time progression includes the time energy within the Planck times in both parts, since overall time progression is unaffected by the quantum nature of time. Time progression benefits from the durations in one part and the jumps in the other and vice versa. It is due to the summation from both halves.

In our TV analogy, there is almost exactly the same sequence. For each glowing pixel[59], the preceding event has already jumped from the previous transmitted pixel, as soon as the new pixel starts to glow. The pixel is now receiving the next signal, moved onwards in time from the previous signal. This is the jump. The pixel now has to glow for a minimum, predetermined period in order to give visual acuity for the frame. Admittedly there are no passing events during this "glowing time" (analogous to Planck time) and the analogy is imperfect in this regard. Finally, there must be blank time, after glowing time, to allow the phosphorescent coating to cool, before receiving the next digital instruction to glow slightly differently, for the next frame. This blank time is simultaneous with the Planck time in the opposite binary part, our pause.

Time does pass during a pause, but the effect, the progress, only comes into being, for the other binary part, from the jump at the end of it, and vice versa. So, in our binary part, we experience time progression over each Planck time, then a gap with nothing happening, then a jump from the other binary part (equal to the progression built up there), then more progression over our next Planck time, and so on. In this way, there is full continuity for the

---

[59] Pixels – The dots on your TV screen which present the colour in that location in the picture at a particular instant.

binary universe, as a whole, without gaps or stutters, but not so for each binary part which is quantised. Until now, we have been looking at our universe through only one eye, the eye of our own binary part, but now we have both eyes open and this new, stereoscopic view clarifies many present-day issues in science, as we shall see later.

By "events" in the above sequence, we refer to either physical movement or temporal progression. In practice, it does not matter which, it is always some combination of both, for all the quanta in the time wave. Again, the combined energy of each adjacent pair of Planck times is invariant.

So, the universe really *is* a series of still shots, just like our TV picture, at least in part. It is the quantum jumps as well as the time quanta themselves that are the deciding factors for driving motion, indeed all changes.

The Two Parts

Each binary part of our universe shares the same void. The two parts are spaced apart, temporally, by up to one Planck time, but there is complete continuity of events, across the binary pair. This being the case, the sum of the adjacent quanta and gap durations must be invariant, to allow each part to mirror the other, at the scale of the wave form. This allows the "negative" to be feasible, producing identical events, and each part has the same laws of nature. Each binary part mirrors the other as an exact "negative" duplicate, in terms of the laws of nature and events in space-time. Thus, we have invariance of the Planck time/pause combinations, or "couples". This is best envisioned as each part "existing" alternately to the other, like two stroboscopes, each out of phase with the other, so only one is flashing at any one time (Planck time). Both stroboscopes "flash" alternately, within each binary part, at 1.855×10^{43} times, every second. It seems that the ancient philosophy of Yin/Yang[60] is fundamentally true, that the universe does indeed have two, identical but opposite parts.

Figure 25 shows how time progresses for the binary pair.

---

[60] yin yang 陰陽 yīnyáng, Chinese philosophical concept – translated as "male-female" "dark-bright", "negative-positive" i.e. two opposites that, together, make a whole

I have included the shaded areas, depicting the density of photons, but the view from our opposite binary part will be physically negative.

The anti-photons will be in the same region as our troughs, existing on the opposite peaks as viewed from our perspective. For the negative world, our "troughs" occur at their peaks.

Note: The vertical axis for the waves is "Time Rate" and the horizontal axis is "Duration" of average, theoretical, constant time.

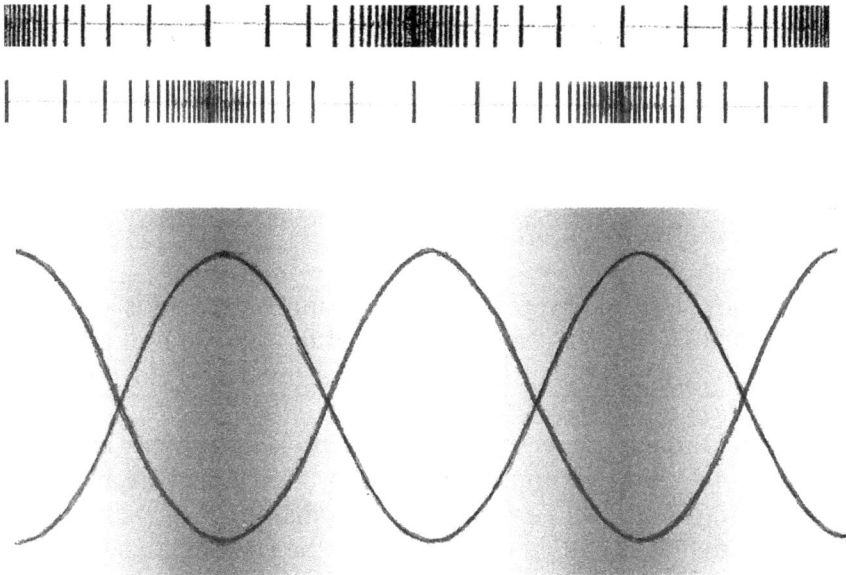

Figure 25: Time in the Binary Universe

The quanta in each part are still spaced in a sinusoidal way, but each is spaced opposite to the other. In this manner, the *"rates of change"* in time are always opposite, or negative, relatively between the two parts, as well as the *"magnitude"* (speed) and *"direction"* (slowing or quickening) of time. This is why the energy of each part is always negative, relative to the energy of the opposite part. The net energy of the pair is always zero, so creation is avoided by this binary energy field

What does this mean for our binary world?

It means that time or change, proceeds in a wavelike manner at the classical scale, but as a step function at the quantum scale. For large slow beings like us, our very existence is governed by the wave and the classical laws of nature apply. But tiny, faster beings "overtake" the wave (they get from A to B quicker than one wave cycle), and so are ruled more by quantum effects and probabilities, by the rules of quantum mechanics. This clearly demonstrates *why* General Relativity is not applicable below the scale of the wave. GR is a *result* of the wave of time, so it makes no sense for entities which experience only fragments of it – particles in the quantum realm.

There is no other view of time that matches accepted science perfectly and that also explains so many present-day, unexplained issues. So, this idea must stand unless, or until, it is disproved, or something better is presented (and I do not believe that will ever happen).

Invariance of the "Quantum Couple"

We will now look at the relative Planck time "durations" in each binary part. We have already established that the sum of adjacent Planck times is a constant, they always must add, to give one Planck time for the pair.

For our binary universe, the overriding requirement is for all energy to be equal and opposite in each binary part, at all times. This maintains the net zero energy at all instances over the wave form and avoids creation. Clearly, this is the case by inspection of the two counteracting time waves in Figure 25. But below the scale of the wave, the quanta are always different. At the trough position of the wave, in one binary part, if the time quantum is zero, then the time quantum in the other part must be one Planck time. At the axis of average time, both quanta are equal at one half Planck time. At the trough position of the other binary part, the quantum is now at zero, whilst the first has increased to one Planck time. We could have added some value to all Planck times, to avoid a zero Planck time at the peaks, but this would mean the binary pair would not have a net zero energy, so we must conclude that the time quanta, in each binary part, oscillate between zero and one Planck time, with the average rate equal to one half Planck time, in each part. In fact, if you take the value at any position in the waves, each wave value is equal and negative to the other. Only a double sine wave has this property of a net zero sum, at all positions on the waves. This is consistent with kinetic energy, in one part, on one of the waves, (say on our wave), being only one half of the time energy of the pair, yet we experience the temporal progression from both waves via jumps and runs.

This "coupling" always provides a total of one Planck time progression for the pair and for any (and every) pair in the wave, but the quanta in each wave are frantically trying to get equalised (and failing), so they oscillate in opposition, perpetually.

We experience our Planck time durations, as well as the jumps from our opposite binary part, so we experience time progression in units of one Planck time. On the macro scale, we only experience the *sum* of the "couples". We cannot detect the pauses, or the jumps, so, our mathematics shows that we experience only Planck time progression.

We can see from the pair of quantum graphs in Figure 25 that where the quanta are the densest, they are at their shortest with the central one becoming zero but, in the opposite graph, the quanta are spread out by the maximum amount and the central one has become one Planck time.

Time progression for the whole binary pair is, therefore, in increments of Planck time.

The Chronon

Many physicists support the idea of a quantised time, or discrete model of time, and the chronon has been suggested as the minimum unit. The chronon, though, is allegedly

> "*a quantization of the evolution in a system along its world line*"

and not a quantisation of time itself. The Planck time is the

> "*theoretical lower-bound on the length of time that could exist between two connected events*".

In short, it is the quickest possible change, in any system.

One might think that the meanings of these two definitions are so close as to be the same, but they are not. The chronon is 6.27×10^{-24} seconds, whereas the Planck time is only 5.39×10^{-44} seconds, in the order of 10^{20} times smaller or shorter than the chronon. It is inconceivable that the minimum period for any change in a system, along its world line, would take 10^{20} times longer than the Planck time.

The chronon is a completely arbitrary quantity, taken as the time it takes light to travel the classical radius of the electron. Clearly this makes no logical sense, in terms of a minimum unit of time. It has no causal justification. The Planck time then, is the obvious choice as the minimum time between events and it has mathematical justification in its place amongst all the Planck units, or natural units, for all the physical phenomena.

11 Challenging the B.U.T.

In any series of tests of an idea against known science, we would expect a false idea to fail pretty quickly. A fairly small number of tests should result in the exposure of some fatal flaw in the reasoning behind the idea. The more tests an idea survives, the more likely it is to be valid. When an idea survives many tests (and when it then goes on to explain problems and issues that are currently inexplicable), then that idea grows in stature and becomes highly likely to be true. Although this is by no means experimental proof of the validity of the idea, it is undeniably supportive of it and, surely, such an idea has merit?

Scientists tend to dismiss anything other than experimental, or mathematical proofs, when it comes to verification of a new theory, yet, hypocritically, they proceed to engage in the same, or similar, logical process for deducing probabilities of outcomes. Examples of this are dark energy and dark matter. There is, as yet, no proof of their reality, but they are nevertheless regarded as a given. The point is that scientists, in general, are predisposed to dismiss speculative thinking as unscientific, unless it is their speculation. In my opinion, they are not always very good at this. Most of the following proposals are made using deductive reasoning and some of the following *is* speculation, but is also deductive, reasonable and certainly useful. On behalf of these ideas, I claim their right to be considered, at least as possibilities.

We will now look at how the B.U.T. fits our current understandings of many aspects of physics, in order to assess the likelihood of its validity. You will find that, not only does it agree with current science (except where I have identified a difference of opinion) but, it also goes farther, explaining some important conundrums that currently defy our understanding.

A Stable Wave

Importantly, the binary universe satisfies the mathematical requirement for two identical, *"opposing"*, standing waves which are necessary to produce a *stable* standing wave of time. This is a well-known aspect of wave theory that the only way to produce a stable standing wave, in any medium, is by the combination of two identical, synchronised, opposing waves and any other type of wave combination is, by nature, unstable. Clearly, the progress of universal time is ultra-stable, otherwise we would not be here.

Antimatter

There is still a fundamental problem in today's physics, regarding the relatively tiny amount of antimatter apparent, compared to the abundance of ordinary matter. The Standard Model requires there to be exactly equal quantities of both matter and antimatter. Physicists are still searching for the elusive, missing antimatter particles, collectively named "Sparticles" or super symmetric particles (SUSYs). Of course, they will never find them. Our binary universe provides an explanation of where all the antimatter "went". In fact, it did not go anywhere, antimatter never was annihilated. It is all around us, today, in exactly the same location as matter, but hidden between our time quanta, in the pauses between the quanta in our binary part.

The positive "stroboscope", of our binary part, only "illuminates" our matter, but the negative "stroboscope", in the other part "illuminates" the antimatter. Antimatter is the matter in the opposite binary part, but we see it as antimatter. Our matter is their antimatter.

This idea also matches, in principle, the Dirac Sea postulate except, we now predict a negative antiparticle for each of our positive ones, as opposed to an infinite "sea" of potential negative particles, with its consequential infinite and therefore impossible, negative charge. In our Binary universe, every particle has an antiparticle. Under normal circumstances, they occupy the same position in space, which is why we will never find them elsewhere. They are hidden from us and only under the extreme conditions created in particle accelerators (or in major cosmic events) can antimatter be exposed as a different, observable entity, appearing to us in a polarised space-time, i.e. from the opposite binary part. Even these antimatter particles are binary in nature and our matter particles are hidden behind each antimatter particle. There is no other way than this for the universe to ensure there are exactly equal quantities of matter and antimatter and so to have avoided creation.

Charge/Parity/Time Reversal Symmetry (CPT Symmetry)

The binary universe cannot permit asymmetry, since this would result in a net creation and creation is impossible. We can deduce, then, that *any* antimatter particles we observe must be normal, matter/antimatter pairs (to maintain symmetry) but in "negative" time, so that we "see" what the opposite binary part would normally see. The existence of the particle, or its local time, has been disrupted and its wave form of time now runs out of phase with ours. Its local space-time is running negatively, just like in our opposite binary half. In other words, its local space-time has become polarised. We can understand, now, that our binary universe is not really two physical entities. It is just the one entity, which is split by the binary nature of time itself. The field of time really is just like two stroboscopic effects alternately "illuminating" everything from the Planck scale up.

Each binary half of a particle is an exact copy of the other, but with opposite energies, in terms of charge, parity and time. Both exist, in the quantum sense in the void, but are hidden from one another, as each half of the universe exists *between* the quanta of time of the other. In this case, time does not actually progress in opposite directions giving a cause and effect problem, but time, in each, simply has an opposite "sense" from the other's perspective. We have seen that the *rate of change* of time rate is always equal and opposite in sense, between the binary parts as well as the *magnitude* (time rate) and *direction* (slowing or quickening). It is this opposite "sense" which produces the relative negative energy of antimatter. Time is the fundamental form of energy, the *only* form of energy. It is time which gives everything its energy and so, "negative" time will result in negative energy.

The binary universe exactly matches the requirement for "Charge, Parity, Time reversal symmetry", or CPT symmetry, at the most fundamental level. CPT symmetry was developed initially by

Julian Schwinger, in 1951, and then further developed by Gerhart Luders and Wolfgang Pauli, in 1954, and proved, independently, by John Stewart Bell (around the same time). In fact, a binary universe predicts this symmetry and even explains why it is so:

> *"Charge, Parity, and Time Reversal Symmetry is a fundamental symmetry of physical laws under the simultaneous transformation of charge conjugation (C), parity transformation (P), and time reversal (T). CPT is the only combination of C, P, and T that is observed to be an exact symmetry of nature, at the fundamental level. The CPT theorem states that CPT symmetry holds for all physical phenomena, or more precisely, that any Lorentz invariant local quantum field theory with a Hermitian Hamiltonian must have CPT symmetry.*
>
> *...The implication of CPT symmetry is that a "mirror-image" of our universe — with all objects having their positions reflected by an arbitrary plane (corresponding to a parity inversion), all momenta reversed (corresponding to a time inversion) and with all matter replaced by antimatter (corresponding to a charge inversion)— would evolve under exactly our physical laws. The CPT transformation turns our universe into its "mirror image" and vice versa. CPT symmetry is recognized to be a fundamental property of physical laws."*
>
> *CPT Symmetry,*
> *Wikipedia 02/06/2017*

The source of this quote was used for simplicity, as it summarises, neatly, the long history and many papers involved in the development of this subject, rather than offer pages of quotes from multiple *more prominent* sources which, when combined, say the same thing.

Now, matter particles are real, positive charge is real and time is a real phenomenon. But, antimatter particles are, also, real, negative charge is real and time inversion is real. The obvious conclusion is that the "mirror image" of our positive universe is just as real as our positive world and that our universe really is a binary universe.

CPT symmetry predicts a binary universe!

What the Binary Universe Theory ("B.U.T.") demonstrates is that ultimately, everything is a function of time. Charge "C" and parity "P" are both functions of time, since time energy is the source of all energy. By the end of this book, I hope you will realise that everything is a function of time and that the energy of time is all there is, the field of energy that allows sequential events or change.

You can see from this description of CPT symmetry that scientists have *almost* stumbled upon the binary universe with its mirror image of our positive part, but they just have not quite made the connection. There is resistance to accepting a real mirror image of our universe, so many have simply ignored this blatant fact and moved on to other things. Some see the negative world, staring them in the face, but refuse to even contemplate its existence. Hopefully, there is now enough substance, here, to demonstrate the Binary Universe Theory.

Quantum Entanglement

There seems to be a conflict between certain rules of quantum mechanics and any causal interpretation of events for entangled particles. This is known as the Einstein, Podolski, Rosen, (EPR) paradox and our new view of time might help to resolve this problem. Ultimately, the paradox comes down to the question, "How can one particle "know" what its other, "paired" particle is doing despite the, potentially, vast distance between them?" There is, indeed, a response by one particle, due to changes the other, entangled particle, undergoes, even over large distances. In the minds of some, this implies a faster than light response, but we should not abandon this universal speed limit. We have no causal explanation for this and we still cannot explain what entanglement is, at the fundamental level.

Entangled systems/sub systems, are clearly "linked" somehow, but no one can explain what this link is, or how there is a correlation between events a great distance apart. Since no information can travel faster than light, it must mean that distance is somehow irrelevant to the relationship between two entangled particles. The only known way this happens, in science (or even in theory), is that the passage of time has ceased between the two particles. Consequently, there is effectively no space, or distance, between them. They both must follow exactly the same wave into the future.

In SR, when time stops (at "c"), distances become irrelevant to the traveller. After all, when your clock has stopped, you travel, anywhere you like, in no time at all. This surely renders distances (in the direction of travel), meaningless. In GR, at the event horizon of a black hole, time has also stopped, distances again become irrelevant to a "resident" of the event horizon, but this time in *all* directions. This gives rise to the "holographic" idea. These two situations demonstrate how the ceasing of the passage of time negates all meaning to dimensions, or to space, and any position in it. So, if we are to apply this knowledge to entanglement, we must accept that, for the "link" between entangled particles, distance has no meaning and this must be because time has ceased, relatively, between them.

This is because, initially, the particles were created at the same moment. They are identical, except for their spin. They are both moving through time at the exact same rate, since one is the "anti" of the other. They move through time on opposite energy waves of the Binary Time field (see Figure 25). Effectively, they occupy the same space, no matter what their separation, which is irrelevant to them. Just like any normal particle in the B.U.T. has its positive aspect and its negative aspect, just one half Planck time behind, the waves of existence being 180 degrees out of phase, each of the entangled pair exists, effectively, in the same place.

Another way of looking at this, is from a total energy point of view.

In the Binary Universe, all matter particles have their exact, negative counterpart in the same space, but lagging behind by one time quantum. This puts them on the opposite time waves 180 degrees apart. So, the link between entangled particles must also be the link between the adjacent Planck times. In fact, these two Planck times are not constant, they vary between one Planck time in duration and zero duration, whilst the antiparticle's Planck time varies between zero and one Planck time. It is the sum of the two that always equals one Planck time and this is the link we are talking about. As the Planck times expand and contract in duration, opposite in sense, this is what produces the two waves of time, the waves of the energy field, like two concertinas expanding and contracting in opposition. See Figure 25 again, for the two bands of time quanta.

It is now clear that the way these Planck times vary are inextricably linked one to the other, positive to negative, such that they always add to exactly one Planck time. So, if you have identical particles but one positive and the other negative, both progressing into the future at *exactly* the same rate, then this situation is identical to our normal particle, with its positive and negative (anti) parts. The link between them is the same.

The point is that the combined energy of the particle pair must remain constant, just like our normal binary particle. If the energy of one of them is changed, then the other has to change, instantly, to maintain the same net energy of the pair, since you cannot create energy. This, I believe, is the fundamental explanation for quantum entanglement and it is enforced by the law against creation.

When electrons and positrons (matter and antimatter) are produced from the same source of energy, then they are "paired" or "entangled". Their charge and spin is always equal and opposite to each other's and if you measure the spin direction of an electron, then the spin of its entangled positron is opposite and is known.

Quantum entanglement might be understood better, when we realise that normal matter, the matter/ antimatter pair is always "linked" together with the respective time quanta spaced *exactly* between the other's time quanta when they are locked together physically in the same space. But particles, where the matter and antimatter parts have been separated and exist on their own, are not possible. There is always the binary field giving us a fundamental duality in everything. Even when we observe antimatter particles they must have their matter particles behind them. So, we can deduce that an antiparticle is still binary, but it has the matter part hidden from us. Its space-time is polarised such that we can observe the anti-world perspective, the reverse of our positive world perspective.

For entangled, matter/ antimatter particles, we are, therefore, advocating complete, conjoined particles, with matter and antimatter in each. It seems that an entangled pair is simply an identical pair, but in relatively opposite, "negative" time, just like the matter/ antimatter within every "normal", single particle. If we change the spin of a normal particle, its conjoined antiparticle also changes its spin, to remain opposite. It is just that we do not normally see it.

If we go back to the sinusoidal variation of time, in the binary pair with the opposite time waves, (Figure 25), we remember that each *pair* of quanta has a constant energy value, but that any (and all) adjacent quanta are always different. It is only their energy *sum* that is invariant. There can only be an *identical* pair of particles, if they have equal but *opposite* energy. The exclusion principle prevents any identical pairs having the same energy, so if they are identical, "numerically", they have to have an opposite sense, to maintain a net zero energy.

The uniqueness of every time quantum is the way all the information about everything in the universe must be "stored" (at any instant). Everything is unique, even if the only difference is the energy *sense.* It is not that information passes between entangled entities faster than the speed of light, nor is it that much information is "known" by these particles, to enable any instantaneous response to the other's experiences. It is more that the binary universe as a whole "knows" everything. It contains all "information" and every time quantum is a unique, "numbered" entity. In this way, everything in the universe could be regarded as "information".

If any change takes place to one of an entangled pair of particles, the universe must instantly "react" and adjust for this change in such a way as to avoid any net creation, even on a temporary basis. The reaction must be instantaneous and, since we are referring to time here, then any such "reaction" must occur "outside of time" and instantaneously. It is the nature of our binary universe that it must maintain equal and opposite, paired particles everywhere, *exactly*, right down to every single quantum, otherwise the symmetry is broken and that is impossible. There is currently speculation about symmetry breaking and this may be possible, but only in situations where the overall, basic symmetry of time energy is maintained behind any apparent physical asymmetry. The universe will not tolerate any inequality between the binary parts, since that would mean a "net creation" and net creation is not allowed. It is impossible.

Since the value of the energy of either binary part is identical to the other's, then any change in one part, must induce the same energy change in the other part, to maintain null creation. If particles are entangled and are identical except for their "sense", then this is a special case, where this universal response becomes visible. This equalising process is going on all the time throughout our universe, but it is only guaranteed to be visible for particle pairs whose only difference is their opposite energy. Any change in one of these paired particles must be matched in the opposite binary' by a change in its "doppelganger". Normally, the doppelganger is the antimatter part of matter and so we do not normally see the response, but where the antimatter particle (or negative pair) is separate and visible, then we *do* see the response.

Now, we do not know if there is a limit to the distance between entangled particles maintaining their entanglement and this cannot be proven by experiment. If there is any limit, then it will be because a better opportunity has presented itself. At some position, the universe might be able to adjust via other means, perhaps with a different particle, or set of particles, and this better opportunity is easier, or preferred, in some way, for the adjustment to happen. Perhaps one particle meets a pair of new ones at the same moment the other particle meets a different pair of new ones. The sum of the energies of each new combination is the same (e.g. our 4 meets a 5 and a 7, but our other 4 meets a 10 and a 2). The sum of each triplet is then 16, both the same. The two triplets might then become entangled. If something happens to one of the original particles, then it does not, necessarily, affect the other original particle but could affect one, or more, of the new ones to maintain equality. It is the *system* that is relevant here, whether the system is made up of one, two, or more particles. We might call this effect "Borrowed Entanglement". This might be a rare occurrence, but it is a theoretical possibility.

The point to note, is the insistence of the binary universe in maintaining a net zero energy and a null creation. Whatever happens, this will always be so, for any (and all) instances and at any instant. It is not so much a deliberate reaction by the universe, but more a self-imposed fundamental restriction on its own behaviour. This is a governing rule of nature, which the universe itself is compelled to obey. It is not that the universe has any intellect, necessarily, but this "obedience" is a direct result of what the universe *is* (and of its constituent parts). There simply is no net energy and none ever was, or can be, created. The universe is incapable of creation, but only of equal, positive and negative, instantaneous "adjustments".

The B.U.T. does give an intuitive reason why, if you change the spin direction of one of a "connected" pair, the spin of the other must also change to maintain conservation, no matter what their physical separation. For this to be the case, we are forced to conclude that, from the perspective of each particle of an entangled pair, space has no meaning and distances are irrelevant. Since they are on exact opposite sides of the binary wave, their times are exactly synchronised and they run together, "holding hands" into the future. They are in tandem in time, there is no relative time rate differential between them, therefore there is no space or distance between them. They are, in effect, "touching".

Here are some pertinent quotes from someone who, until recently, was a research scientist at the Perimeter Institute for theoretical physics:

> *"The precise definition of entanglement is that two or more objects share the same wave function."*
>
> *"In the grand scheme of things, everything in the universe shares the same wave function."*
>
> *"Everything is still "touching"".*
>
> *Fotini Markopoulou*

I believe I have given here the best, perhaps only, intuitive explanation of quantum entanglement, so far, because knowledge of our binary universe leads me to a different way of thinking. So again, the B.U.T. is strengthened. I cannot offer you a more detailed explanation of the causality, but it demonstrates that it is inevitable in the binary universe. The physical mechanism is still a mystery, since it involves the notion of relationships "outside of time" and, until we can get our heads around this aspect of our universe, it will still seem to be:

"A spooky action at a distance"

Albert Einstein

One last point, remember that time is the only entity in the void. There *is* nothing else and time is the *"giver"* of causality, it is not bound by it, so we should not expect causality to be a requirement for any changes which are limited to the phenomenon, within the phenomenon of time and with no effect on the "outside" world. A responsive change in one of an entangled pair is limited to a change in time only, there is no "external", physical impact on the universe. Causality simply does not come into the action. Thus, such changes in time can indeed occur, instantly, in these circumstances.

Energy Conservation in the Binary pair

Our universe came from nothing and today, it still amounts to a net nothing. This idea is accepted today and presented to the public by the likes of Dr. Lawrence Krauss, a physicist, a popular science presenter and author. The universe must have a net zero energy, otherwise it could never have emerged from nothing. The golden rule of nature is that you cannot create and so the original zero energy must be maintained throughout the process of "creation" and over the lifetime of the universe. We have seen, in the preceding section, how the binary universe theory predicts such a situation and even explains, logically, how this occurs.

We also see from the B.U.T. that, for every positive particle, there must be an exact, negative counterpart, a discovery made by Paul Dirac from the mathematics of quantum mechanics. He discovered that QM mathematics predicted an antiparticle for every matter particle and scientists still search for these "Sparticles". According to the B.U.T. these "Sparticles" are hidden behind every particle, existing on the negative wave. It does not take too much thought, to realise this is the only possible arrangement that can ensure the net energy of the universe is maintained at zero, at all scales and for all time.

Our Binary Universe maintains the principle of conservation of energy, the natural law that dictates we can neither create, nor destroy, energy. The complete, twin universe is a net zero energy system since everything is balanced with a perfect symmetry, from the very moment of creation and until the end of time itself. This symmetry is quite perfect because it is a digital symmetry. Any "creation" is always balanced by an equal and opposite "creation" in the other binary part and likewise, so is any annihilation. In this way, all energy is conserved at a universal net zero at any instant on the time waves.

The following equation from Chapter 8 describes the total energy in any closed system, including each part of the binary universe, since the universe is the ultimate "closed" system:

$$E_{total} = \sum \left[mc^2 - E(\Delta t_m{}') + \frac{1}{2}mv^2 - E(\Delta t_{KE}{}') \right] = 0$$

Expressed in simple English, this means "The total energy of the system equals all the mass energy minus the gravitational time dilation energy from that mass, plus all the kinetic energy minus the inertial time dilation energy from that motion". This sum total of energy always equals zero, since the time dilation energies from mass and motion are exactly equal and negative to the mass energies and kinetic energies. Thus, the net energy of the universe is always zero, the possible exception being the vacuum energy, which seems at first glance to be wasted. This will be explained soon.

With our new understanding of the transference of time energy, from the field, into kinetic energy, with increasing motion, the above equation can now be seen to reflect this relationship between the energies of time and of speed. We also understand that all mass takes time energy from the field, by virtue of its internal momentum and so the equation makes perfect sense. The equation is applicable to both binary parts, the net energy within each part is zero and the overall, total energy of both parts is also zero. It is not necessary to consider other forms of energy, since these are either potential (and are therefore not yet in operation) and/or they ultimately boil down to (or ultimately become) kinetic energy (e.g. temperature or heat is the kinetic energy of vibrating molecules).

The time dilation energies are always negative and evolve from the opposite binary wave. This also applies from the perspective of our opposite, negative part, so we can re write the equation as:

$$E_{TOTAL} = \sum_{A,B} \left[mc^2 + \frac{1}{2}mv^2 \right] = 0$$

where A and B represent the two binary parts.

It is now clear that no net energy has ever been "created" and that any energy, in one part, is always mirrored by the numerically equal, negative energy, in the other. Thus, we have avoided creation and the symmetry is perfect! (Almost).

The Vacuum Catastrophe

Many physicists believe that "The vacuum holds the key to a full understanding of nature", but there is a problem. The cosmological constant problem. One of the greatest unsolved mysteries in physics, refers to our current inability to establish the energy of the vacuum, the zero-point energy. Richard Feynman and John Wheeler calculated zero-point energy to be an order of magnitude greater than nuclear energy, with one teacup full having enough energy to boil the Earth's oceans. Clearly this is wrong and astronomically so. It is now thought that this calculated, large amount of energy is somehow almost cancelled by something else as yet unknown, to a very low, residual value.

Using the upper limit of the cosmological constant, since this is a real observation, the calculated value of the vacuum energy is about 10^{-9} joules per cubic metre, a minute amount. However, quantum electrodynamics and other theories require it to have a value of the order of 10^{113} joules per cubic metre and this huge discrepancy is referred to as the "Vacuum Catastrophe".

I have proposed that the vacuum energy is simply a reflection of the energy of time passing in the void. It is the only energy there is, after all. I ask myself how this can be, when the two time waves are equal and opposite and cancel themselves leaving a zero energy field? But is it a zero energy field? Do the waves cancel exactly? If you look at Figure 25 and the binary time waves, they look equal and opposite, but they are not (quite). Since the energy quanta follow each other, as they alternate between positive and negative, then depending on your point of view and in which wave you exist, you might regard the other wave as lagging behind your wave by up to one Planck time, or perhaps leading, rather than lagging. Either way, there will be an energy imbalance of one half Planck time, on average, between the waves for each wave cycle. The Planck times, experienced by each half, vary between zero to one Planck time, with the average being one half Planck time, over one complete cycle.

This being the case, it suggests that there is a "positive" vacuum energy of one half Planck time, when observed from one wave but an opposite, "negative" vacuum energy of minus one half Planck time, when observed from the other wave. There is vacuum energy and anti-vacuum energy. Only in this way can the net energy of the universe, the net energy of the slightly out of phase time waves, remain at exactly zero for the binary pair and so, yet again, avoid creation.

Only with a binary universe can we have a vacuum energy in both halves, yet maintain a net zero vacuum energy for the pair. This unique property of the binary energy waves provides even more support for the B.U.T. and resolves the vacuum catastrophe.

The two waves almost eliminate each other, except for the tiny vacuum energy value we observe, in practice.

The invariance of the speed of light?

We will now address the issue of photons travelling through a constant frequency, *standing* wave, whilst appearing as different frequencies of light. If we consider two slightly different coloured beams of light, the wavelengths are different and this is what we see as the different colours. This effect applies to the whole of the electromagnetic spectrum, with wavelengths ranging from radio waves of the length of a football pitch, through infrared, visible light, ultra violet, up through X-Rays and ultimately at the highest frequencies, or shortest wavelengths, high energy Gamma rays, with wavelengths almost down to the scale of subatomic particles. Gravitational effects apart, the question then arises, "How can we have different "colours" or wave lengths of electromagnetic radiation if all the photons travel through the same time rate field which has a constant frequency?"

There is only one possible answer. Electromagnetic radiation must travel through the vacuum, within a tiny range of velocities *just below* the theoretical speed of light. The different frequencies of radiation must be the result of photons travelling at different speeds through the cycles of time. If our understanding of the wave nature of time is correct, then at the speed of time (speed of light), radiation would be infinitely *red shifted,* when viewed from the stationary frame and there would be no wave, no light.

From this, I deduce that no photon at all can ever travel at the speed of light, but only slightly slower. So, for the highest speed photons, the light will appear initially as very long wavelength radio waves. As photon speed slows further, it becomes increasingly less *red shifted* and so effectively *blue shifted* throughout the whole of the electromagnetic spectrum until, at the slowest speeds, we finally observe high energy gamma rays. Photons cannot exist at speeds much slower than light speed and the electromagnetic spectrum fades to nothing, at the high end of its possible frequency range.

It is easiest to consider a particular photon, at the very peak of the time wave (or peak of the light wave) and see what happens to it, as it progresses through the time rate field at a speed marginally slower than light speed. As it moves through space and as the volume of the space progresses through time, in a sinusoidal way, then for slower speeds than light speed, (slower than the time wave), the static cycle of time will hit its peak sooner over shorter distances during the linear progress of the slower photon. So, the "peak" photon, and therefore all the others around it, will peak earlier and over a shorter distance travelled. The distances between the peaks will reduce, as speed slows and the wave form of light will become compressed. The light will become *blue shifted* for speeds slower than light speed. This is consistent with working backwards (slowing down) in Figure 23.

This view of time and light means that the highest energy light, with the highest frequencies, should travel at the slowest speed. Consequentially, although this effect is only marginal, the higher frequencies of EMR should take measurably longer to reach us, from very distant objects, and this tiny effect should be detectable, over very long journeys. This may seem the wrong way around to conventional thinking, so let us just check how this happens. The nature of time is that it speeds up, then slows down, cyclically, in a sinusoidal way, at any *fixed point* in space. In this sense, the sinusoidal wave of time is a *standing* wave. We need to remember though that it is the whole volume of (local) space that undergoes this cyclical variation of time in unison (ignoring gravitational effects for the moment).

It may seem counter-intuitive that the higher energy light is, in fact, slower but we need to understand that it is the cycles of time which impart cyclical energy to the light, rather than the light having its own kinetic energy, due to its speed. Photons have no mass after all. Certainly, the slower photon has less "momentum" over average time, but the time wave now has more opportunity to excite the photon, as it moves more slowly through the time rate field. Remember, it is the wave form of time that gives particles their wave functions and the fields of photons their wave nature. As high frequency light hits our detector, it is cycling much faster than any lower frequency, redder light and, in the same way that electrical current has more energy with higher frequency, so does the light.

From this understanding, I predict that high energy Gamma rays travel the slowest of all the types of EMR and that they should, therefore, take slightly longer to reach us, than longer wavelengths. Crucially, there is experimental evidence which confirms this, but some quite ridiculous and very creative explanations are being tabled by mainstream physicists in order to try and explain it..

There have been anomalies, as far back as 1987, which question the constancy of the speed of light. A study on the 1987 supernova, SN1987A, showed that the visible light took 4.7 hours longer to get to us than the associated neutrino burst. This demonstrates that different wavelengths/energies of light do, indeed, travel at different speeds. James Franson, of the University of Maryland, suggests this slowing down of light is due to the polarisation of the vacuum, but this mechanism is complex and not yet understood. It is, as yet, unproven. He is guessing, because he has no idea of the causality.

More recent measurements were taken of the light pulses from a feeding black hole, by the "Major Atmospheric Gamma Ray Imaging Cherenkov" (MAGIC) telescope. In 2005. Astronomers from the MAGIC team pointed the telescope at a Blazar, within galaxy Makarian 501, which is half a billion light years distant. The results showed that the higher energy Gamma radiation, emitted from the Blazar, took four minutes longer to reach us, than the lower energy gamma radiation, over the distance of 500 million light years.

Mainstream thinking is that this slowing down of higher frequency light might be due to the effects of the "quantum foam" on the path of the radiation, but the quantum foam is, again, still an unproven postulate. This proposed mechanism of resistance to motion is *highly* speculative. Again, they are guessing, since they cannot explain these results satisfactorily. The B.U.T. sees the cause as light, naturally travelling at all the speeds available to it, depending on its energy. These results provide important evidence in support of the proposed new theory of time, especially since the effects of vacuum polarisation and quantum foam are speculation without any causal explanation.

If a photon travels through the wave form of time, at a speed slightly slower than the speed of light, it will have a wave function with a higher frequency and will appear *blue shifted.* The frequency of the photon is increased in the time rate field in which we exist and, as photons travel slower and slower, they will eventually disappear altogether. In effect, they will not exist. They *cannot* exist at speeds much slower than light speed. This suggests that photons are, by nature, some effect within (and of) the time rate field, rather than "constructed" from some other, fundamentally different entity. (This is further surmised a little later in this chapter). The slowest particles are, therefore, gamma ray photons, at the extreme "Blue" end of the electromagnetic spectrum. High energy gamma rays have a wavelength of around 10^{-20} metres. At smaller scales than this, is the world of subatomic particles. It is because of the difference in the nature of time, itself, at these different scales, that we have only radiation above the scale of electromagnetism and only subatomic particles below this scale. Again, this suggests that both EMR and particles are merely different forms of the same thing, at different scales in the field of time. It is becoming more and more clear that particles are made from energy, the energy from the field of time.

Implicit, in our new understanding of time, is that the fastest photons are at the red end and the slowest are at the blue end of the EMR spectrum. I had already arrived at this conclusion by the above deductive reasoning, before I became aware of the recent observations of distant gamma ray bursts and the 1987 supernova results. These observations confirm this order of arrival, so give even greater credence to the B.U.T.'s sinusoidal, *standing* wave of time. These observations are experimental evidence of this *standing* wave. The mainstream, however, has an expectation, a preconception that Gamma rays should arrive *before* the longer wavelengths of EMR, simply because they are more energetic, but they are still struggling to understand why this was wrong.

Critically, whether the mainstream accepts this yet or not, these recent observations of Gamma rays from distant galaxies demonstrate that different frequencies of light (different energies) do, indeed, travel at slightly different speeds and that the speed of light is not the exact constant we have believed it to be, until now.

> *It is the speed of TIME that is invariant, NOT the speed of LIGHT,*

but even the speed of time is not constant across distance as this varies with the strength of gravity and so all we *can* say is that:

> *"UNIVERSAL TIME" is invariant,*

but, yet again, even universal time is not precisely constant.

I am proposing that time, itself, is speeding up, very slightly, over the eons. Even so, the concept of a constant universal time is a good local approximation, a good starting point for understanding the nature of light and time at the macro scale.

So, the speed of electromagnetic radiation covers a tiny range of speeds, just below what we, currently, believe the theoretical value of light speed to be. That the propagation of electromagnetic radiation is not, precisely, invariant, is a startling revelation, since it appears to contradict Relativity's assertion that all photons must travel at exactly the "speed of light". However, this range, this variation in the speeds of photons is miniscule, although it is enough to produce this effect. For all practical purposes, the invariance of the speed of light can be taken as true. Albert Einstein was correct, for his day, but we now have to move on, in the face of our new definition of time and this indisputable, supporting evidence.

The road to Quantum Gravity

We have established that, when an object accelerates, the time quanta transfer their temporal duties, from moving its mass through time, into inertial duties in moving the object through space. Time quanta happen everywhere throughout space. Time is a phenomenon, within the void, so we must understand that a "stationary" mass uses most of the local time quanta, to move it through time and the remainder to move it through space, in the sense that certain subatomic particles which make up the matter have some spin or rotation and therefore, momentum. Otherwise, it just sits there and moves through time.

So, we *also* understand that to move a stationary, larger and/or denser mass through space, at the atomic scale (atomic spin), will require more of the local quantum energy, compared to that needed to move a stationary, smaller/less dense mass through space. This being the case, it is not surprising that local time becomes slower in the region of the mass and more so, the larger or denser the mass (more spinning particles in the space or volume). The mass is taking up some of the finite number of time quanta to move it through space, because of all the momentum energy within the mass (compared to a region with no mass in it at all).

The result is that fewer time quanta are left to move the region through time, so time slows down. This creates the local reduction of the time rate field at the surface and, consequently, the surrounding time curvature, as the field recovers, with distance above the surface. This is the cause and effect of mass induced time dilation, the curvature of time and, therefore, for Newtonian gravitation. This internal energy, within all matter, drawn from the field of time, is the fundamental cause for *all* gravitation in both weak and strong fields. There is no other known causality.

In the extreme, a black hole will use *all* the available, local time quanta to feed its enormous spin energy, so none are left to move the rest of the local universe through time. It "eats" all of the time and its effect, on time, is equivalent to travelling at the speed of light, yet another confirmation of the equivalence between the energy of time progression and kinetic energy. From this we can say that:

> *The energy of the time dilation (in one binary part) = the energy of the motion producing it*

Since we are advocating that gravitation is purely the result of time dilation/curvature and, from our understanding of work done, is equal to force times distance moved, then this is the same as saying:

> *Gravitational "force" = Acceleration force*

They are "equivalent". Clearly, the first statement is a more fundamental version of the second, so we have now interpreted the Equivalence Principal in *the* most fundamental terms. This is *why* the Equivalence Principle is part of the laws of nature.

Some scientists like to envision a gravitational field as space, itself, "falling in", or accelerating, towards the centre of the mass. For a black hole, space might be said to be falling at the speed of light, on the event horizon. From this view, again, we might also see *why* the equivalence principle is valid. The force exerted by gravity on a stationary individual, is equivalent to being accelerated upwards, in space, by the same acceleration value as "g", for the large body or planet. One could also view this as space accelerating downwards, to give the same effect. The *practical* meaning of the equivalence principle, is that the time curvature will attract you or "force" you, but, now, we can again see this happens, regardless of whether the time curvature is from a gravitational field, or from your inertial acceleration. It is the time curvature that is the mechanism of gravitation. This is the same as the time curvature of your acceleration upwards. Remember, it is the *gradient* of the time curvature that defines your gravitational attraction, as well as the numerical value at any point in the field and so it is with acceleration. It is the rate of change of your velocity (the slope) which exerts an acceleration force, not your actual speed at any instant.

In less extreme cases, for entities less dense than black holes (like planets), only *some* of the local time quanta are used to move the mass through space (rotation and spin or momentum of all the subatomic particles). In empty space, time simply passes, without impediment, at the rate of universal time, (1.855×10^{43} Planck time per universal second). There are no other demands for its energy. But, when a star or a planet is using some of the local time quanta, to move it through space, then less are available to move the local universe through time, so time slows down, in the vicinity of the mass. This is *why* the sum of kinetic energy, for both binary parts, is numerically equal, but opposite (in sense), to the time dilation energy it creates, i.e. the gravitational energy of the field. This statement is equivalent to Einstein's field equations which will say the same thing, if we view them as suggested in Chapter 7.

This transference, of the duties of the time quanta, is the cause and effect of mass induced time dilation and of time curvature in the field. But, we can see there is no direct causality for any real distortion or curvature of space itself, this effective space curvature is a result of the way the time rate changes, over distance. The inertial time dilation effects, from all the spinning, subatomic particles, all contribute to the effect. This mechanism of time dilation is the direct cause of time curvature, within the field, and therefore for Newtonian gravitation. On top of this pure Newtonian effect, the varying field of time "distorts" velocity and acceleration, *relatively,* within the field and this is the effect predicted by General Relativity (and explained at the beginning of Chapter 7). It is equivalent to the abstract curvature of space. Again, space, the void, is ultimately nothing. It is not a field. It has no directional properties, so it cannot influence motion. The abstract distortion of space is unreal. It is merely an equivalent effect, caused by the distortion of events from the time curvature.

Of course, the curvature of time does not come to an abrupt end at the surface of any mass (say the Earth), but continues outwards, decaying in all directions, towards infinity, in a spherical way. This is verifiable and measurable as the Earth's time dilation field. It is the *only measurable* entity within any "gravitational" field, so it is proven by "experiment". The gravitational field, however, has not been proven to exist, it is merely *implied*, from the behaviour of the particles within it, yet, this behaviour can now be completely explained, by the time dilation field and its distortion effects. I hope I have convinced you that:

> *The gravitational field is a figment of our creative imaginations!*

It is no wonder that we cannot find the graviton. If the field does not exist, then its particle certainly does not exist either, so I predict they will never find the graviton. Only the field of time exists and this must have its own particle, but this particle can only reflect the field, not its curvature. The particle cannot be the force carrier of gravity, although it is indirectly related to it.

The Lorentz transformation reflects the quantum cause of inertial time dilation. It is the link between quantum effects and time curvature and, therefore, ultimately with gravitation. The Lorentz transformation is important, since this is the way time dilates with speed and, therefore, with the way the number of "skipped" quanta increases with speed. Critically, a similar equation will also describe the number of skipped quanta in a gravitational field, since time "curvature" is the cause of both inertial time dilation and gravitational time dilation, due to the equivalence principle.

The Lorentz transformation, or some form of it, is applicable to both inertial time dilation and gravitational time dilation.

From this logic, we can see there must be a mathematical relationship between Lorentz and Planck units. This will form part of the gateway to the mathematics of quantum gravity, not just the incorporation of GR into QM, as in "Loop Quantum Gravity" and String theory. This relationship must, systematically, reduce the number of time quanta experienced, or conversely, increase the number of those skipped, with increasing speed. This will happen slowly, at first, but at an increasing rate, so that, finally, no time quanta are experienced by the moving entity.

It does not take too much insight to realise that, for any velocity v, then:

$$t' = \frac{1}{\gamma} = \sqrt{1 - \frac{v^2}{c^2}} = 1 - \frac{N_T}{N_L}$$

Where t' is the proper time rate for the moving entity as a proportion of universal time, N_T is the number of time quanta skipped and N_L is the number of Planck lengths travelled.

Checking the sensibility of this equation, we see that at the speed of light, the time rate t' becomes zero, as the fraction of the numbers of time and length quanta becomes 1. For low speeds, t' approaches unity, (i.e. 100% of universal time) and the quanta fraction approaches zero. This equation enables us to calculate the *exact* number of time quanta which are skipped, over any particular distance and at any particular speed, relative to the *standing* wave of time.

Significantly, it is by this effect that we can now determine how fast an object is moving, both relative to us and, also, relative to the *standing* wave of time, (the preferred reference frame). We can then work out how fast *we* are moving through this field and by how much our time is already dilated from universal time. This accurate, digital information will become essential for any journey at very high speeds and over vast distances, namely for interstellar travel.

For gravitational systems the time rate at any radius r is given by:

$$t' = \sqrt{1 - \frac{2GM}{rc^2}}$$

Also, our previous energy equation is:

$$E_{total} = \sum \left[mc^2 - E(\Delta t_m') + \frac{1}{2}mv^2 - E(\Delta t_{KE}') \right] = 0$$

This equation shows that the total, net energy of the universe equals zero and is the summation of the static mass energy, or time energy, of all the mass, minus the energy of the time dilation, caused by this mass, plus all the kinetic energy of motion at the macro scale, minus the energy of the time dilation caused by this motion.

So,

$$E_{total} = \sum \left[Mc^2 - E_{\sqrt{1-\frac{2GM}{rc^2}}} + \frac{1}{2}mv^2 - E_{\sqrt{1-\frac{v^2}{c^2}}} \right] = 0$$

This applies to all closed systems so, in very general terms, it describes the energy within any single gravitational field.

I am not suggesting this formula is, in any way, rigorous enough to produce a theory of quantum gravity, but I make the point that, ultimately, gravitational energy can be quantified as the number of Planck times used for speed, instead of temporal progression, within the volume of the field. Quantifying gravity is therefore, effectively, a counting exercise.

A mathematically rigorous version of this equation will be difficult to formulate. It would need to determine the gravitational energy due to time dilation, in a spherical field that reaches to infinity, it would also need to define, accurately, the kinetic energy due to the motion of all the spinning particles, within the mass itself, again expressed as a spherical distribution. The total energy of this field equals the total momentum energy of the particles making up the mass.

This understanding shows us that gravity is purely the curvature of time around the mass, which is caused by the internal kinetic energy of the mass using up some of the local time energy from the internal time rate field, plus the external, temporal "tension" effects. This effect accumulates throughout the mass and causes the time rate, at the surface, to drop. Above the surface, there is a resistance to any sudden change in time rate over small distances, so the time rate increases only gradually, as we move away from the mass. Let us see if we can understand exactly how this resistance to sudden time rate change happens, this temporal "tension".

Looking, again, at the binary time waves in Figure 25 and considering that they operate in every fixed position, say around the surface of a large mass, then we must ask, "What effect would this *red shifted* surface wave have on a *non-shifted* wave say one millimetre above the surface?"

Well, we know that each "quantum couple" must add up to exactly one Planck time and there is no escaping this relationship. That being said, we know in a gravitational field that time is *red shifted* at the surface and less so as we move away or increase our elevation. The only way this can happen, for a quantum effect, is the same as for inertial time dilation. Only complete quanta, or quantum couples, can be eliminated from the field, as the field *red shifts.* So, we must conclude that *only* complete quantum couples can be regained by the field, as the field *blue shifts* higher up in the field.

The driver here, is that the field always tries to run at universal time, energy has to be taken from the field to reduce it. Higher up in the field, from the large mass, there is no energy being taken locally from the field, so the field tries to run at universal time. The only reason it cannot do so, is that the time wave directly below it is *red shifted,* so the wave we are considering can only be different by one quantum couple, for every Planck length gain in elevation. In this way, there is a limit to the recovery rate that can be gained with elevation, so this is why the field "curves" in accordance with the Schwarzschild metric, as shown in Chapter 7, Figure 15.

I believe I have now explained, in completely logical terms, the cause and effect of gravity and at this point, I am forced to leave it to the mathematicians to interpret these ideas and express them in the language of mathematics.

But, *to be clear*, they will have to dispense with the notions of discrete units of length, or shapes, and all things geometric and replace them with equivalent effects, from the energy of time in order to make headway.

Inertia

From our new knowledge, of the way time quanta are transferred from temporal duties to inertial duties with increasing speed, we can understand that, when a mass changes its speed by a certain amount of energy, then a certain, related number of quanta are diverted from temporal duties to inertial duties. At the quantum level, this is not just a one-off transfer, since the increased speed requires a continual, additional supply of energy, to provide the continual quantised acceleration over each Planck time.

Since the inertia of an object is proportional to its mass and, therefore, to its use of time quanta for inertial duties (spin energy), then we can deduce that the number of quanta transferred to inertial duties, with increasing speed, is dependent upon the number of quanta already used for spin. Since mass does relatively change with speed, we can also deduce that the number of time quanta, being transferred to inertial duties at any instant, is dependent upon the increasing number of quanta already being used for inertial duties, at that moment. Effectively, this is *why* mass increases with speed, relatively.

As an object accelerates, it has to draw more time quanta (energy), at a faster rate, from the number of quanta occurring over time, within its own volume. It uses more quanta, per second, from those happening, per second, because it is moving faster. Ultimately, at the speed of light, it uses all the time quanta at the same rate as they occur (within its own volume). So, in this situation:

- Its clock has stopped, because there are no longer any time quanta available to move it through time,

- *Effectively,* its mass has become infinite, because it cannot be accelerated further. There are no time quanta left, to transfer to the energy of motion.

    The inertia of an object = A measure of its resistance to acceleration.

We can see that its increasing inertia, "the difficulty with which it can be accelerated", is determined by the increasing number of additional time quanta being transferred to kinetic energy over time. When there are none left, its inertia is infinite, relative to a frame using all the time quanta for temporal duties (relative to the stationary frame – the field of time). If there are many quanta not being used for motion, such as when it is "stationary", then its inertia is proportional to the number of Planck times it is using to feed its spin energy. This proportionality is applicable to all speeds which are insignificant, compared to the speed of light. It is obvious that for a mass using, say, twice the number of Planck times for spin, its mass and inertia is doubled and that the number of these Planck times defines the spin energy, defines the number of spinning particles, defines the number of atoms, defines the mass and inertia. It has twice the inertia.

The initial energy needed to accelerate is dependent only upon the spin energy within the matter, but the additional energy required as it accelerates is also due to the additional quanta which have been transferred to inertial duties. This increases in accordance with Lorentz, because the increase in relativistic mass is caused by the inertial time dilation, caused by the increase in number of quanta used for speed.

$$m_{rel} = \frac{m_0}{\sqrt{1 - v^2 / c^2}} = \gamma m_0$$

But, the greater the mass, the more spinning particles there are and the greater the number of time quanta already used for spin. So, a lesser number of time quanta are available to convert to inertial duties. A lesser number are available for time dilation, so this is why anything with mass can never reach the speed of light. There are insufficient time quanta available to transfer to inertial duties, to get to light-speed. To get to light speed, we need *all* the time quanta that occur, to transfer to inertial duties. We need a massless particle with no internal spin momentum. If some quanta are already used for spin, then there is a shortfall. This gives us a new understanding of the factor γ:

$$\gamma = \frac{1}{\sqrt{1 - \frac{v^2}{c^2}}} = \frac{1}{1 - \frac{N_{SPIN} + N_{INERTIA}}{N_{AVAILABLE}}}$$

Because,

$$m_{rel} = \frac{m_0}{1 - \dfrac{N_{SPIN} + N_{INERTIA}}{N_{AVAILABLE}}}$$

Where:

- m_{rel} is the relativistic mass,

- m_0 is the rest mass,

- N_{SPIN} is the number of time quanta per second used for spin. This is a measure of m_0 and is fixed.

- $N_{INERTIA}$ is the number of time quanta per second used for inertial duties. This is variable, dependent upon velocity.

- $N_{AVAILABLE}$ is the total number of time quanta available. This is fixed at the rate of 1.855×10^{43} Planck time per second of universal time.

(I have used one second as the period over which these numbers occur, but I could have used any period and applied it to all of them).

The equation tells us that gamma can only ever reach unity for a massless particle. For anything with mass, it will never quite be equal to 1, since N_{SPIN} has some value and the denominator is thus, marginally, less then unity. We can also see, from the equation, when our speed has increased (so that the sum of inertial and spin quanta equals the available quanta and all the time quanta have been used up), that the denominator becomes zero and the relativistic mass becomes infinite. For anything with mass or spin, this happens *before* we reach the speed of light, so anything with mass can never reach the speed of light. For slow speeds, the number of inertial quanta is practically undetectable (Lorentz) and the inertia is defined, largely, by the spin quanta, the energy of its spinning particles, the mass. Gamma, of course, practically equals unity, unless the mass-energy becomes huge. For a black hole, the number of spin quanta equals the available quanta and gamma has reached infinity, without any motion of translation at all.

One could deduce that black holes have infinite inertia and cannot be accelerated, so they should be fixed at their original velocity on their formation in the stationary time rate field.

To continue, any inertial mass increase is only relative to the stationary frame, not the moving frame. Within the moving frame, the time being experienced is dictated by the diminishing number of time quanta, time is slowing down as speed increases. The spin, however, is still defined by the number of spin quanta, which is fixed in the moving frame. Inertia remains constant and is unaffected by speed. In other words, inertial mass increase is purely relative. Again, the cause of any and all effects from the motion is the inertial time dilation, not any real increase in mass and certainly not length contraction.

We can also see that, for entities with zero mass (photons), then there are no quanta used for spin, so all the time quanta are available to be converted to inertial duties – can be used to dilate time, so photons travel at the speed of light (more or less.)

Inertia is the energy required to change, (increase or decrease) the rate of transfer of time quanta, between temporal duties and inertial duties.

But why does it take energy to change the rate of quanta transfer between duties? Surely a simple transference of energy should require no input at all! We need to look at what is happening, for the complete binary pair, to get the whole picture. If we are using more quanta for kinetic energy and less for temporal progress, then we are "extracting" temporal energy from the universe (the field of time), to use as work (kinetic energy) and this time energy has to be split two ways, with each kinetic half going to opposite halves of the binary pair. One way of looking at this is that our binary, part for example, "loses" half the time energy, in the process of converting the quanta, but the other binary half has to gain the same amount. Energy must be transferred, to "gift" this same amount of kinetic energy to the other binary part. It is this demand for the transference of energy, to the other binary part, which requires us to exert a force, to input energy. Without a binary universe, no such force would be necessary to accelerate objects, since the process would only entail the transference of quantum energy between duties, not between binary parts. To transfer energy within our binary part costs nothing, but to give energy to the other part requires us to input it. If there were no other binary part, then no energy at all would be required to produce motion, and the whole universe would instantly disappear amidst infinite chaos. In other words, a binary universe is essential for it to sensibly work at all.

This aspect of a binary time slows everything down, from an impossible, unrestricted causality to a manageable, finite rate, which allows the universe to function. This property of a binary time field suggests it is the Higgs field, since the Higgs field does have this same dampening effect on events. The Higgs gives mass to matter, as well as having a small, non-zero value throughout the vacuum. This is further speculated, a little later, in the section on "the time rate field".

Perspective

The wave nature of time, is the limit beyond which we cannot see clearly. We cannot see very well, beyond the scale of the wave nature of time and closer in to the quantum range of time. Sinusoidal time is the environment within which we exist. This sinusoidal form defines all our observations and physical laws. The wave form of time defines all the classical laws of nature.

We live in, and on, the sea of time, the wavelike ocean of time. We sit offshore, bobbing up and down and we look at the beach, wondering why it goes up and down relative to us. We wonder why everything has a wave function. We wonder why items floating close to the shore (the quantum realm), bob up and down differently to us and why their rules of behaviour are different to ours. We marvel at this non-intuitive behaviour, convincing ourselves we will never be able to grasp what is happening.

Below our wavelike time field, faster than our wave, where events occur so fast they are not dictated by the wave, we see different rules of behaviour than our classical, wave driven rules.

Our observations become distorted, fragmented and less predictable. This is because we are observing a scale where our sinusoidal time wave is larger (slower) than our focus and, increasingly, we lose sight of our time wave and the associated classical laws of physics, as we focus down towards the quantum scale.

At this scale, the time *quanta* start to become more significant, compared to the time *wave* until, ultimately, at the Planck scale (right at the shoreline), there are only the Planck time quanta and no time wave effect at all. At this scale there is no time and so, surprisingly, there is no space either and everything could be said to be "touching".

This idea has been researched by Fotini Markopoulou[61], while at the Perimeter Institute for Fundamental Research, in Canada. Markopoulou agrees that you *can* envision a universe without space, but not one without time. Time is more fundamental than space and space emerges from time, as I have shown here, independently.

At the Planck scale, there is no space or time!

Scientists are still trying to apply General Relativity, at the fast scale, in the quantum environment, but GR cannot be applicable, in this realm, since time curvature is the slow scale change in frequency of the time wave, over many cycles.

How can anything that exists, below the scale of the wave, faster than the wave and in less than one wave cycle, be influenced by any curvature, over many cycles? That is impossible. The only way we can reconcile GR with QM, is to understand this and accept that their rules of behaviour operate in different realms. GR really does break down in the quantum domain, so it can never be reconciled with QM. They are different sets of rules for different situations.

---

[61] Sadly, Markopoulou has since changed career direction, a great loss to Science.

Uncertainty

Below the scale of electromagnetic radiation, the rules of quantum mechanics start to apply. The quantum world is a world of wave functions and probabilities. Everything has a wave function, but until now, no one really knew why. Everything has a wave function, because everything moves through the wave of time, even if it is only to sit there and move through time, itself.

It is unclear, from our observations, whether a travelling entity is a wave of energy, or a particle, or sometimes one and then the other but, if we consider the proposed wave nature of time, then things become a little clearer. It becomes evident that photons are particles at *all* stages of their journey, but that our observations, at the small scale, do not permit us to see events, accurately, from within our own sinusoidal time.

As a particle moves through the field, fractionally slower than the speed of light, its temporal position "slips" behind our static wave form of time and we cannot predict when it will arrive at a particular location, or where it will be at a particular point in our time. This is because its speed (and therefore momentum) cannot be defined from within our field, since its relative velocity varies, depending on the point it is at on its own time wave. This is not constant, nor does it follow the same form as our time wave. It is not, simply, that it is variable, but that it also varies in a different cycle to ours, with a different wavelength. As a result, since its predicted position depends upon its speed, we cannot define its position at any point in time, on our time wave. We cannot know exactly *when* (on our clock), it is *where*! Because of this lack of a clear observation, from our static sinusoidal time field, even by mathematical prediction, if we focus on exactly where a particle is, we lose track of how fast it is moving and vice versa. If we observe its velocity, we lose focus on its position. At any moment, we can only "observe" one phenomenon, at a time, between fields of time, with different sinusoidal waves. We are thus driven to make predictions only from a probability analysis between two wave forms of different wavelengths, which are out of phase. This is the "intuitive", physical/temporal cause and effect for the Heisenberg Uncertainty Principle.

I imagine two small boats on a rough sea. I am on one of the boats, which is sinking, and I am trying to get onboard the other one. My boat is bobbing up and down differently to the other, so it is very difficult to know when, or where, to grasp the ladder. If I focus on a particular position and wait for the ladder to get there, I will not know how fast the handle position will move past me and I could injure my hand, as it is snatched by the moving handle. If I focus on the speed of the handle, which is varying and with a different cycle to my boat, I am unable to predict where it will be at any future moment. When I grab it, I am likely to grab it in the wrong place and miss. This might seem an impossible problem. But there is a way to make predictions, if only we know the wave functions of both boats, or the combined wave function of one, as viewed from the other.

The only way to break this complex problem down and to be able to make predictions is by using probability analysis. For this, all you need to know are the wave properties, or functions, of the wave-like behaviours.

For light, most of the photons are clustered around the peak of the time wave, with the highest probability of there being one at the peak. The lowest probability, of there being a photon, is at the trough of the wave.

The nature of statistical analyses means that is all you need to know, to accurately predict the probability of an occurrence of a photon. It is the same with our boat analogy. If you know the amplitude and frequency, or wavelength, of each boat's oscillations, then you can work out the probability of any particular rung, on the ladder, being in a particular position, at a particular time. For the boat, the best you can do is to maximise the probability of the rung being within your grasp but, of course, that may not be enough to save you. You would need to try to grab it many times, to increase your chances of one successful grab. For the photons, there are already many occurrences of photons, at any one time and so there are already many chances of finding a photon in the wave, at a particular position.

This is why quantum mechanics is so accurate at making predictions. It predicts the *probability* of an occurrence over many events, rather than particular outcomes. The probabilities are born out with incredible accuracy.

Length contraction and the edge of the universe

Our observations, of the light speed entity, will show that its length is contracted to zero-length (by the time dilation of motion). From the reverse perspective, observations from the moving frame with its stopped clock, again shows lengths as contracted. Even the Earth will look like a flattened disc, from the moving perspective, but it is clear we are not really flattened, within our frame of reference. This is the meaning of the statement, within Special Relativity that, "proper lengths never change". The relative *lengths*, as observed from each frame, are indeed both contracted by the same amount, they are compressed to zero-length, in the direction of relative motion and, as far as length is concerned, all motion is, indeed, purely relative. This is why the mainstream gets it wrong! They focus on the dimensions and the geometry, but seem incapable of grasping the temporal aspects.

Because of the real differences in time rates, observations of time related events, like speed, cannot be symmetrical. From the light speed frame, with a zero time rate, you will observe the complete life of the universe, until the end of time, in an instant, just like a photon. The static universe's clock will look infinitely fast and, relatively, it will actually *be* infinitely fast, as predicted by Albert Einstein in 1905, (although he was perhaps unaware), and as verified by Hafele & Keating in 1971.

You may ask,

> "How can this be the case, over different distances?" "If you see the end of time, in an instant, then how can you stop after several light years travelled and see the time elapsed as only several years?"

To understand this, you need to realise that your clock really has *stopped,* in absolute terms and the static universe carries on, in time, but at a finite rate. As you travel, distance becomes irrelevant to you, only becoming relevant again, on your deceleration. You might travel five light years and five years *will* have passed on your return to the "static" universe, but you will have arrived instantaneously. You could have carried on and travelled, say, ten light years and ten years will have passed, in the "static" frame, on your "return", but you would still have arrived instantaneously, and so on, for *any* distance travelled, even to the edge of the universe.

Muons again

Muons are created by cosmic rays impacting on atomic nuclei on entering upper layers of the Earth's atmosphere. The muons produced by these impacts continue in, roughly, the same direction as the cosmic rays and they head towards the Earth's surface at a significant fraction of light-speed. Muons are very short-lived particles and we know their lifespan (c. 2.2 micro seconds) but, as we track them through the atmosphere, they travel much further than their natural rate of decay should allow. They should decay before reaching the Earth's surface, but more muons than predicted actually reach the surface. This is seen as an illustration of time dilation *and* length contraction, where the muon's existence is much slower than ours due to its slow "clock", so the distance travelled appears much shorter to the muon. However, we have already shown that the Earth is not flattened by a light speed entity passing close by, so why should we believe that the distance to the Earth's surface is shortened, in reality. *Effectively,* it is only shortened for the muon due to the inertial time dilation, the only real effect of motion.

Actually, this *"effective"* (but imaginary), length contraction is an effect of the time dilation, meaning that the full distance is covered, but in a shorter time for the muon, due to the muon's slower time rate. It is not necessary to invoke *both* time dilation *and* length contraction, since only *one* of these is necessary to explain the increased distance travelled by the muon, in its short lifetime. The static, proper distance passes by faster for the muon, because of the relative *blue shift* of the stationary frame. It covers the real, non-contracted distance, faster, due to the increased speed, relative to the Earth's "static" frame. This effect is *not* due to any real contraction of distance. The Muon clearly travels the same distance as we see, but, within its own frame, it covers it faster than the time we measure. This is indicative of *faster than light* travel but only within the moving frame. This is purely due to the time dilation.

There are claims of experimental results showing actual pictures of egg-shaped or "pan-caked" muons demonstrating length contraction, but this is just a relative illusion. It is like observing an object, through an optical lens, but in this case the lens is the lens of time. As light speed is approached, the muon also approaches a situation where its time will stop, therefore the back of the muon must "arrive" at the same time (proper time) as the front of the muon and, effectively, it must compress to zero-length, relatively, from our perspective. But, Special Relativity states that proper lengths remain unchanged, within the moving frame. This means that whatever change in length we might observe, from our frame of reference, *must* be an illusion since there can be no contraction in the frame of the muon. It is only the time dilation that is real. Therefore, this reality must result in the increased speed in the muon's frame, *not* in any length contraction.

The mainstream does not accept the notion of different speeds in each frame, since it cannot accept super luminal speed, even though it is purely relative and limited to the moving frame. So, it holds the velocity in both frames equal and allows the resulting length contraction. But the time dilation is *real* and the length contraction merely a relative illusion. Time dilation is the only available cause for any effect. Length contraction can never be a cause for anything, since it is unreal. This means length contraction cannot be presented as the reason for the different distances covered in the same times, as measured by both the more accelerated and less accelerated clocks. It is the time dilation which should be the governing factor in the deductive reasoning within Special Relativity, not length contraction. Traditional SR does not allow different speeds in each frame, neither can it accept the notion of infinite speed in the moving frame, so it has rejected the interpretation of SR given here. This is despite the acceptance of a zero time rate in the frame of the light speed entity. This is a contradiction within the mainstream interpretation of Special Relativity. Mainstream thinking has been prevented from considering this view, because of its fear of superluminal speed (relative), and a preferred reference frame. They have allowed fear to dictate their thinking, which is not scientific.

Because of the reality of the time dilation and its effect on relative velocity, the new theory maintains that the current mainstream interpretation of Special Relativity is quite simply, wrong!

It may be argued that this is not science and is mere opinion, but, then, so is the mainstream understanding of a common velocity between frames, an opinion!

The geometry dictates *either* length contraction *or* acceleration due to time dilation and that only *one* of them must occur. We know which one from Hafele & Keating. We must make a choice, since only one of these is necessary to explain all the effects. I just happen to make a different choice, based on experimental results. I am unafraid of the consequences of my choice. Some will disagree and will continue to do so until my proposed experiment is carried out.

We now have sensitive instruments able to detect the transverse shifts of inertial time dilation, as observed from both frames in relative motion. We have space vehicles fast enough to produce a detectable level of inertial time dilation (Transverse Shift). There is no longer any excuse for not doing this simple test as a priority, if only to silence "alternative speculation". I challenge the mainstream to prove me wrong. In reality, of course, this experiment has already been carried out, indirectly, by Hafele & Keating. The results demonstrate that time passed quicker in the stationary frame, than in the moving frame. My prediction is merely that this will be visible in a direct measurement experiment, between frames with measurable time rate differences.

So, a Muon's time really is slower, but its rate of existence cannot be slowed down, just because it is moving relative to us, or for that matter, relative to anything else. What does the muon care about *our* relative velocity? Nothing! There can be no effect from *our* "stationary" frame which causes the existence of the muon to progress at a slower temporal pace.

Nevertheless, there has to be a real cause which makes the muon's clock slow down, so what is it? Exactly what *can* possibly slow down time for the muon? SR shows that time dilation *must* occur with motion, but it does not provide any causality. Since the muon is moving through space-time, then, it can only be some aspect of space-time that causes this time dilation. Once more, since space itself has no properties on the macro scale, then we must look to the phenomenon of time in our search for this cause.

As previously explained, both the wave and quantum nature of time *will* have this effect. Since the speed of the muon is close to light speed, then it is passing through the time rate field, through the cycles of time, almost as fast as the time wave occurs and the muon's time will become *red shifted*. The wave form of static time will be significantly *blue shifted,* relative to the muon. That is to say, the time rate for the static universe will be *blue shifted* from the muon's perspective. This matches the fact that the muon's proper time rate is *red shifted,* relative to the static universe, simply due to its speed through the time rate field.

> *The time rate field is the preferred reference frame, the standing wave of time.*

So far, we have considered only an isotropic or uniform time rate field, but we know that time passes, at different rates, everywhere in the universe depending on, not just your relative speed, but also on your elevation within gravitational fields. In regions of space, with measurable gravitational fields, the cycles of time have a longer wavelength relative to those in weaker gravitational potentials. This is due to the slowed time rates, close to massive objects. It is demonstrated by the *red shift* of light, when looking into a gravitational field. The longer wavelengths, at low gravitational potentials, are due to a stretching of the wave form of time, which extends each wave of light, so we get *red shift* as well as Newtonian gravitational attraction, due to the time curvature within the field. This explains why gravity and time dilation are always coincident, since they are one and the same thing. We also note, again, that there is no tangible evidence for any *other* cause of gravitational attraction, apart from the time curvature.

One might argue that we cannot call the time rate field *a* "frame of reference", since it is in fact an infinite number of frames of reference, as the field varies in energy throughout space. So, we must understand what we mean, in practice, by the term "preferred reference frame". The preferred reference frame is defined by the value of the time rate, at the location of the event we are considering. It is the local *standing* wave of time and is quantified by the following:

- The net time rate at a particular elevation, or position, within all gravitational fields

- The time rate at a particular velocity, relative to the temporal *standing* wave.

In practice it is always,

- Some combination of the two.

So, contrary to current scientific belief, there *is* a preferred reference frame, or field – a field against which all speed must be measured. Unlike some mainstream physicists, I do not panic over the implications of this, but look forward to discovering how we might change our thinking, in order to accept it and marry it with other aspects of relativity.

Electromagnetism

Electromagnetic radiation is a field of radiating photons (with their anti-photons), but the anti-photons are in the dark bands between the light bands in figure 25. This field has both an electric and a magnetic component. The wave forms of these components are said to be in different, abstract, "planes" and we depict them as being ninety degrees rotated from each other, in the usual pictorial presentations. Our other binary part runs temporally negative, relative to us, and this will be the reason why the anti-photon has a different effect on us from "our" photon. It still uses time energy, but they are physically separate from the photon. This must also result in a loss of time energy, in our positive world. The field of photons and anti-photons consist of two waves which, when added together, gives a plain field of a constant number of these positive and negative photons. This is why we get the magnetic wave peaking *between* the peaks of the positive wave.

So, we might deduce that we experience the electric field from the photons and the magnetic field from the anti-photons. In the other binary part, these effects will of course be reversed.

We have established the cause of the gravitational field as time curvature in the field. Time energy is taken from the field by the subatomic spinning particles in the mass, for them to use as kinetic energy. This works two ways, of course, with both binary time waves losing energy quanta. So, it must be the same with photons.

Photons do have a minimal mass, or at least a momentum, so they, too, must draw some energy from the field, albeit a minimal amount. Incidentally, we might conclude from this that photons can never quite reach the speed of light, or the speed of time, and we have already seen that this is, indeed, the case. Both photons and their anti-photons draw time quanta and produce local time dilation, in both binary parts. We know what the effects of a time dilation field are, from our knowledge of gravity, so we should expect to see similar effects from EMR. Indeed, we do see such effects.

We see the direct electrical effects from our photons and we see an, apparent, "action at a distance" from the anti-photons, the magnetic field.

Why electric effects from our part and magnetic effects from the other part? Why not the other way around?

Well, we know that the gravitational effects, in our part, are induced from the time dilation in both parts. There is an apparent "action at a distance" and magnetism is the same type of effect. I, therefore, select the magnetic field as being the local, strong time curvature effects from the time dilation.

Magnetism is the same effect, fundamentally, as gravitational time curvature, but on a very localised scale. It could be said that magnetism is the "near field" time curvature, whereas gravity is the "far field" effect. Clearly, the electrical effects are those generated from the photons within our binary part, since they are direct effects.

The Multiverse

How might all this fit with the "Multiverse" idea? The multiverse is a postulate claiming that the physical void is infinitely vast and so, inevitably, there will be an infinite number of other universes formed. The logic, here, is based on probability. If our universe happened (and it did), then it must have been possible for it to do so.

If it was possible, however unlikely, then there was (and still is) some degree of probability associated with the event. It had a finite probability of it coming into existence, so there is also the same or similar probability of another universe coming into existence somewhere else in the infinite void. This being the case, then another is possible and another and so on, ad infinitum. Thus, there must be an infinite number of universes within an infinite void.

The false assumption, here, is that time would allow this unlikely event to happen, but time, itself, is a phenomenon associated with our universe, alone, and time is the phenomenon which restricts unlikely "possibilities" from happening. We will see later, that the *beginning* was the beginning of time and *time* is the fundamental causality for everything, for the whole universe. Where there is time, there is our universe and there can be only one of it.

I am proposing that our universe was created by the beginning of time. Time defines the universe, in which case, time does not happen in any void beyond our universe because there *is* no void there. There *is* no "there". Where there is no time, there is no space. The only requirement for the creation of our universe was the starting of the cosmic clock, the beginning of time. Everything else is a consequence of this, from the creation of the most fundamental particles, in the early universe, to the formation of the first stars, through to the galaxies of star systems we see today. I remain sceptical of the multiverse idea. It simply does not make sense and it certainly goes against the fundamentals of the Binary Universe Theory. The B.U.T. is supported by logical deduction and mathematics, but the multiverse is pure speculation.

Mainly, we are concerned with our universe, just *one* of the possible universes of any multiverse. There is no guarantee that any of the others, if they exist, will be similar to ours. So, I have no intention of ever attempting to understand any of the other universal possibilities. They are irrelevant, surely one universe is more than enough!

Super Dooper Pooper Symmetry

The requirement for a perfect symmetry, at least for the first few moments of the Big Bang, has driven the scientific community to search for this symmetry, especially in view of the apparent, gross imbalance between matter and antimatter. Remember, with a binary universe there is absolutely no such imbalance at any stage of the proceedings, whatever our observations show us.

Today, it is believed that soon after the Big Bang, matter and antimatter annihilated each other and that the matter we see now is simply the remnants from this annihilation, due to the initial inequality of these constituents. Frankly, this argument seems more than a little "flaky". We have conjured it up, in our attempt to explain the apparent lack of symmetry. Its basic flaw, is that it presupposes an initial inequality, in the first place, which implies a "creation" from nothing and we know this is impossible. Symmetry is an essential property of nature, at all times, from "creation" to the present day. This view is also naïve, in that it is based upon our present-day experience of how matter and antimatter annihilate each other. It is blind to the possibility of them existing in the same space. This mistaken idea leaves us with an asymmetrical universe, composed mainly of matter and within which we now, paradoxically, search for symmetry, yet we know there must be an overriding symmetry at all times.

Several proposals exist for how this symmetry might be achieved. The recently proposed theory of "Super symmetry" now suggests that there is a separate partner particle, for every fundamental particle in the current version of QM. (The B.U.T., independently, makes the same claim.) Comically, these partner particles have been collectively named "Sparticles". Of course, this would double the number of fundamental particles and the search for these is ongoing. As of writing, no evidence has been found for these extra particles and there is already some disappointment that, contrary to expectations, none have yet been discovered. Clearly, the B.U.T. provides just such a super symmetry since the antimatter "sparticles" exist in the same space as matter particles and so this proposal must stand, unless or until the first "sparticle" is discovered, separately. I think we will wait a very long time for that.

A perfect, unbroken symmetry is an essential element in a Binary Universe, not just for the first few moments after the Big Bang, but also since then, for all times up to the current epoch and then on, for all eternity. All matter is balanced with its antimatter pairing and symmetry is maintained by the "creation" and annihilation of matter/ antimatter in matched pairs only.

This perfect symmetry encompasses not just particles, but the very phenomenon of time, itself, from which the particles must have emerged. The wave form of time, in each binary part, is always symmetric to the opposite wave form in the other part. They are a reflection of each other. The B.U.T. explains how the symmetry of everything is perfect, down to the minutest detail, even to the extent of tiny "out of phase" of the time waves, the dark energy in the vacuum.

A Very Speculative Overview of Quantum Field Theory

Quantum field theory proposes a separate field for each type of particle, a photon field for photons, a quark field for quarks and so on. There are many fields envisioned within quantum field theory, but the B.U.T. proposes just the one field, the time rate field from which all particles can emerge depending on what is "done" to the field and the different properties of the field, for these different "impacts". The theory proposes that the time rate field is the one field from which *all* particles have been created. This field does have some energy, even where it is basically flat and devoid of particles since time always passes everywhere and at a very slightly increased rate over time, so there is a latent, energy in the vacuum.

But what is a particle? When we think of a particle, we might try to use our everyday experience of the macro world and so we think of a speck of dust, or a molecule, or atom of gas, but this is a completely inappropriate view of particles, at the small scale.

Fundamental particles, the building blocks of matter, cannot be things such as dust or gas molecules or atoms or even quarks because those things, themselves, are made from the very fundamental entities that we are trying to envision. They must be something else, of a completely different and fundamental nature. We must go back to the first few moments, after the Big Bang (if there was one), to understand what particles must be.

At this time, when matter or physical particles or gas did not yet exist, then the only entities we can envision are the void, (which is nothing) and the passage of time. Do not forget time *is* energy. Time must have existed before anything else could happen, since events needed to unfold. So time was, *verifiably,* in operation to allow the creation of matter. We must conclude that the first fundamental particles were formed from space-time and, since we are proposing that the empty void (space) is simply nothing, then the only entity available from which to create fundamental particles was the phenomenon of time itself, the field of time.

Fundamental particles *can only be* very small, local effects of the passage of time, limited by the discrete ways in which a wavelike time can take effect within three-dimensional space, at the very small scale. This *must* be the case, since there is only space and time, from which to create anything. Different particles must be the result of the limited number of ways that time can behave, at the smallest scales. They are ultra-stable, perpetual "micro vortices" of time curvature, whose internal behaviour and external effects on the field, result in the properties of the different particles we observe from experiments and that we determine by Quantum Mechanical theory. We know that some particles are not stable and have very limited life spans, but their existence is nevertheless possible, if only briefly.

In this thinking, we might be heading towards the view of fundamental particles being composed of various combinations of micro black holes orbiting other micro black holes but, admittedly, this is very speculative. Nevertheless, black holes *are* permanent since their time is stopped relative to the macro world, so they are perpetual by nature. They are the only things, we know of, that actually do last "forever", apart from the time wave, itself. The available combinations of (or possibilities for) these black holes may, actually, give rise to the different fundamental particles that exist, or can exist. These different combinations are equal in number to the six extra dimensions in string theory and, it seems to me, these six possibilities of dimensions are, in fact, the six possibilities for the way in which time energy (and therefore space) can "curl up" to form stable particles. After all, there is no such thing as space without time, so if we are thinking of dimensions, we are ultimately thinking of time.

If we consider the changing time rate, over the eons, and how time is measurably faster now, than at the moment of creation, then we might suspect that particles can be created from the field of time, perhaps only during a period with a certain, specific and slower time rate than our current time rate or, perhaps, during the phase of inflation, when time sped up from zero to somewhere near its current rate. This process of creation would have been extremely rapid and intense, with particles "flooding", or condensing, into existence from the field of time. As the time rate increased, beyond a certain rate, and, as the process of inflation drew to a close, then creation of fundamental particles, from the field became impossible and the formation of our universe was complete.

Perhaps, we can still detect the simmering remnants of creation as virtual matter/ antimatter pairs flitting in and out of existence, within the vacuum. When we create the very hot conditions during collisions in particle accelerators, we create very small volumes with intense, high energies.

These extreme concentrations of energy will result in a significant, localised slowing of time, due to gravitational effects, because energy equals mass, which produces time dilation. It is perhaps this slowing down of time, to a rate similar to that in the early universe that again allows certain particles to be "created", but only ever in matched pairs of matter/antimatter and always with a perfect symmetry.

The Time Rate Field

A crude analogy of the field of time might be the surface of an ocean of fluid, where we can create small vortices or whirlpools on the local scale, but where the ocean surface on the macro scale can only ever be wave like. Particles are analogous to whirlpools and the time rate field is analogous to the ocean surface. It is by this analogy that, perhaps, we get a glimpse of why there are only particles below a certain scale and only EMR above this scale, since vortices are only possible at the small scale, compared to the vast ocean surface. Then on the next scale, the next plane, there are the large macro curvatures superimposed on the waves and these are the "tides" of *gravity!*

We can now start to see that gravity is not created, directly, by the interaction of particles, but is the direct result of changes in the time rate over distance, of time curvature on the large macro scale. It is the "tides" in the ocean of time. It might be interesting to consider that the viscosity of the ocean's fluid is quite relevant at small scales, but much less so at larger scales and this varying effect, with scale, might be a useful analogy for the effects of time being different (quantised) at the small scale and of it being wave like at the scale of EMR.

Every fundamental particle would simply be a perpetual disturbance of the time rate field and would, of necessity, possess a wave nature, being a disturbance of the wave form of time within the field, having an effect on the field around it, which, itself, is wavelike. Clearly, because of the finite duration/energy of the Planck time, there will be only discrete possibilities within which particles exist, since time curvature (and that is all you can do with time) will only be possible below a certain scale, under certain conditions, at certain sizes, with certain momentums, with certain combinations and complexities, with certain resultant properties and with the essential stability, at least for long lived particles. I suspect that the mathematicians will have a "field day" with this idea (no pun intended) and perhaps the nature of every type of fundamental particle can be predicted, mathematically from this initial assumption.

Of course, whatever disturbances (particles) can exist within the time rate field in our binary part, there will, of necessity, be an associated disturbance in the other part, having opposite energy, in exactly the same location and we can now start to "see" how the binary universe operates. Each fundamental particle is a "swirling" effect of the time quanta, an extreme, complex localised time curvature, which will be duplicated in the opposite binary part, but with a one half Planck time separation and with a relative, negative sense. In this way, no matter was ever "created" except as an equal and opposite pair of disturbances for any entity or group of entities, in fact *all* entities. Of course, the next question is "How did the binary time rate field come into existence?" The next chapter makes a proposal for this.

Since the time rate field is the field which creates gravitation, by virtue of its own curvature, at the large macro scale, (tides in the ocean), then it will be the field which is closest associated with gravity. This fits perfectly with the idea of gravity as the curvature of time, with Newtonian gravity. It may be quite possible to explain this mathematically, by inventing virtual particles (gravitons), in order to comply with the known rules of QM and this is what we seem to have done. Nevertheless, it is the *field* that is the active entity, the cause of the effect of gravitation and *not* the particle which can be produced as a result of some local "impact" on the field. The existence of a particle is less important than the fact that it proves the existence of its field. Critically, although we may be able to create particles from the time rate field, it does not seem possible to create a particle from the *curvature* of the field and curvature is the cause of gravity, not the actual field.

The Higgs field is understood to be the closest associated with gravity and it also has a non-zero, average value. It has energy, or it *is* energy. It gives mass, or inertia, to matter and slows an, otherwise, chaotic universe down to a manageable speed. We have already surmised that a binary, quantum-like time field is what gives mass to matter. It also slows things down and avoids a runaway self-destruction of the universe. The properties of the Higgs field are the clearest indication, yet, of the time rate field, the field from which everything evolved. The Higgs Boson though, is not the graviton. It seems it is the particle that emerges from the field of time, but the field is not the cause of gravitation. It is the *curvature* of the field that causes gravity. I therefore deduce, there is no such particle as the graviton and this would explain why no one has been able to find it.

Mathematical analyses of the possible disturbances and their probabilistic outcome (life span and properties) should be based on an understanding of the nature of quantised time, rather than from geometric manipulations of a hypothetical curved space. We can see how the application of GR, a geometric interpretation of space-time, becomes quite inappropriate at this small/fast scale. The quantum foam is simply a demonstration of the unpredictable nature of space-time, at the very small scale, where there is no wave and the classical rules of nature do not apply.

Such things are fascinating ideas, but if I go any further I will be speculating beyond reason and I may have already done so. Speculation apart, this new idea of a quantised yet wavelike time, across a binary universe, unifies quantum field theory into one field, Einstein's unified field. It also gives clear, causal explanations for the wave nature of light and inertial time dilation and gives us a subtly different understanding of the quantum world which directs us toward a real "Theory of Everything".

And Finally ... the Beauty Contest!

I did not set out to "manufacture" an attractive theory, or to seek to identify beautiful ideas. The theory has evolved naturally and of its own volition, simply by applying deductive reasoning and strict causality to accepted science. Strangely, it is accepted amongst the mainstream that, if a theory is a beautiful one, it is more likely to be valid than a, relatively, more ugly one. I find this amusing, but, nonetheless, I do understand why. The human mind is a part of our universe and aesthetic appreciation is a part of the mind. By our very nature, if something seems beautiful to us, even an idea or series of linked ideas, then there is a certain justification in our giving some weight to our perception of its beauty. So, "Is the B.U.T. a beautiful theory?" To my mind, there can be nothing more elegant than the perfect symmetry which is necessary for the Binary Universe. The symmetry is between the binary parts and its extent is all encompassing, right down to any and all individual Planck times. What is more, this perfection will have existed since the beginning of time and it will last until the end of it. It is an eternal perfection. I cannot envision a more beautiful idea, than the construct of a binary universe and any alternative proposition must necessarily be more complex, less comprehensible and less beautiful. There are now new questions to pose, like why is time quantised and binary, how did time begin, how does time relate to the Big Bang and was there really a Big Bang? On and on we will go, searching for truth amongst what we believe to be facts. Our most formidable enemy, in this quest, is our own misplaced stubbornness, our resistance to changing our thinking. As always, new knowledge brings new questions, but this is as far as I should go for the moment since, in the practical sense, we have solved the riddle of time.

12 The Big "Fizz"

I am not a professional physicist, an expert in mathematics, or a particle physicist and so I am unable to construct a theory of the beginning, working backwards in time towards the Big Bang, as is currently being attempted in mainstream science. I can only present my offerings from a higher level. I am forced to start at the beginning and deduce my way forward, in time, from some initial condition. I must start with nothing and work my way forward. However, unlike current mainstream physicists, I have an understanding of our binary universe to guide me along the way and this will certainly direct my thinking, differently. It may allow me to progress understanding, where others are stuck with philosophical issues. This constructive approach has certain advantages, given that neither approach can ever be proven in full. Perhaps both approaches will always be necessary, for us to get the whole picture but, I admit, the following text in this chapter is very speculative in nature.

On the other hand, it can do no harm to speculate, so here goes.

In the beginning, or rather, "just before" the beginning, there was nothing. There must have been nothing at this point, since if there were something, then the beginning would already have happened in order to have produced that something.

So, our first step is to try and understand the very concept of absolute nothingness and to see where this initial condition leads us.

We live in a world devoid of nothingness. Everywhere we look there is something. Even the vacuum of outer space is not absolutely empty and it has a tiny energy value, the vacuum energy. Our brains are simply not wired to envision nothingness and this places a limitation on our imaginations. Nevertheless, we must try and look beyond our present environment and from outside our personal perspectives, if we are ever to understand the beginning of everything.

When we say there was "nothing" we must also accept there was no time. I have identified the phenomenon of time to be energy, a physical field of energy and so this also cannot exist until or after the beginning. In fact, I have also proposed that all matter particles are simply localised "arrangements" of the energy of the field, with their corresponding extra, curled up "dimensions" and this explains the equivalence of matter and energy at the fundamental level. So, from this understanding, we see that first, we must create the field, before any particles can evolve from it. This field of energy, of time, must come first, so we deduce that the beginning must have been the beginning of time, the beginning of the wave like field of energy. Every particle, every entity, somehow evolved from this field, at some later stage. Initially, in the absence of matter, this singularity would have had zero mass and not the infinite density predicted by the mainstream's reverse extrapolations.

We have also seen that, without time, there are no dimensions, no space, so this situation defines the initial condition of our singularity, an empty universe without time and without size, in other words, absolute nothingness.

The universe, (let us call it that), before the beginning must have been a singularity with no time, no volume, no matter, no energy, no mass, no gravity, no temperature and so on. Such a "singularity" then is not infinitely dense, since there is no energy or matter present and no gravity either, since gravity is the curvature of time caused by the presence of matter.

So how do we get our field of energy from absolutely nothing? Well, we must consider that, although there is nothing physical in existence, there was, nevertheless, one rule at this stage – that any (and all) virtual events (possibilities) must follow – this is the rule *against* creation. You cannot create something from nothing. Yet, there is just one process which allows a particular form of "creation". You *can* create an equal positive and negative energy at the same time (and in the same place), thereby maintaining a net zero energy of "creation". Only in this way, is it possible to get something from nothing, but it always has to be this "binary" something, with a fundamental duality in everything. Everything must have its duplicate of opposite energy.

There are indications of this today from the emergence of virtual matter and antimatter particle pairs, flitting in and out of existence in the vacuum.

There is one more rule in this situation – that there *are* no rules. The rules of nature have not yet formed, since nature, itself, is yet to be defined. In this situation, the result is that *anything* can happen. Anything is allowed, so there are infinite possibilities. The fact that anything is allowed and all such random events are possible (however unlikely), is another way of saying that "Anything that is not forbidden is inevitable", in other words, "Murphy's Law". You may find this somewhat comical but there is a scientific understanding about Murphy's Law. In quantum physics all possibilities (quantum states) exist, until the wave function is said to collapse, leaving only the one reality from all these quantum states, or possible realities.

So, if we apply this idea to our, randomly occurring, virtual positive/negative quantum energy pairs, our virtual binary quanta, flitting in and out of existence "within" our singularity, then all possible arrangements of these quantum fluctuations are possible and so they do all happen, even the very unlikely ones. But, there is no pattern to them. They are completely random and unstable and they never repeat. They never continue, so are doomed to remain only virtual by nature, just mere possibilities. There is one exception, the very rare, highly unlikely, pattern of a binary energy wave. Now, since all possibilities, all arrangements of these quantum fluctuations *will* happen, then so will our binary energy wave happen. It does not matter how unlikely this event is, or "how long" we have to wait for it to occur, since time is not yet passing and probability is an aspect of time. When time does not pass, then you might say, everything will happen all at once or, as is often said, "Time is the process which prevents everything from happening, all at once". So, in the absence of time, everything (all possibilities), happens all at once.

The only virtual events that are important though, "after" all possibilities have happened, are those possibilities that endure. The most important possibility is the one that is compelled to repeat the production of its virtual pairs of quanta in its own, unique pattern. Within our singularity, all possible events are compressed into, what to us, is one single moment in time, just like dimensions are also compressed into one single physical point in space, from our present-day perspective. So, our binary energy wave must occur within the singularity. It is inevitable!

There is something very different about our binary energy wave, compared to all the other possibilities. This binary energy wave *is* continuous by its very nature. In itself, it provides the environment for change, the phenomenon of time, and it is this effect of time that gives the field its continuity. The field provides its own continuum! All other possibilities are unstable and die a natural death, instantly collapsing back into nothingness, just like the virtual particle pairs in today's vacuum. Our binary energy wave though, is self-perpetuating. Each wave is driven by the other and each relies on the other to produce the next linked quantum in the series. Any one quantum defines the next quantum and so on. These series of events cannot deviate from this relationship. They are compelled to continue, to repeat in a wavelike manner. We see that, by its very nature, our binary energy wave forms a continuum, *our* continuum! Today, we understand and experience this energy wave, as time. Everything has evolved from it, lives within it and is driven into the "future" by it, changed into the next quantum state, by each quantum pair or couple.

Every fundamental, physical particle can *only be* some local "arrangement" of these energy quanta, so the wavelike field must give up some of its quanta for them to do different things within a particle, rather than contribute to the wave, locally. Each of these "arrangements" is a stable, localised series of events, in which one quanta follows the previous quanta, within the locality of the particle. Then, at larger, slower scales within the field, all entities are carried along with the wave, stepping forward one quantum pair, at a time, into the future, into the next, sequential state of existence. This is how matter must have evolved from our initial field of energy. It is the only possible way. This is why the number of extra "dimensions" in string theory is the same number of fundamental stable particles in the standard model. They are the different possible ways that the energy of time (and therefore dimensions) can be wrapped up, locally, to form these perpetual, fundamental particles. We can also "see" that any such arrangement must take its energy from the energy field, so the field of time must diminish, where there are particles, it must *red shift,* where there is mass. This is the fundamental causality for mass induced, or gravitational, time dilation and therefore, for gravity itself. It is the energy reduction, in the field of time, feeding the particles in any mass that causes gravitational time dilation and its associated phenomenon, Newtonian gravity.

There is a simultaneous effect from the commencing of this wave. We have already established what happens when the clock starts to run. Space emerges from time. This relationship between time and space is evident from both Special and General Relativity. In SR, as we approach light speed, lengths in the direction of travel start to contract until, in the theoretical limit, at "c", lengths in the forward direction become effectively zero. The stopping of time in the moving frame causes lengths to become meaningless. Then, as we slow down from light speed, time starts to pass again. Dimensions in the forward direction regain their meaning, so lengths, or distances, emerge again. Space re-emerges, as time re-emerges.

In GR, the same effect is evident. As we approach the event horizon of a black hole, time slows down, until, at this horizon, time stands still. Space has shrunk in volume, increasingly, on our approach to the horizon. Here again, lengths become effectively zero, but this time in all directions and space becomes a spherical hologram around the surface of the horizon. Again, working backwards, we see that as we move back out, away from the event horizon, time starts to pass again and space "expands" from zero volume, as a result. Space emerges as a result of the commencing and increasing time rate. General relativity shows that space compresses further in and relatively expands again, further out. We can say that the passage of time creates lengths, distances, volume and space. Time "creates" space or, better, that space "emerges" from the phenomenon of time. You cannot have space without time, but time is fundamental whilst space is emergent. Time is the cause and space is the effect.

So, when our binary energy wave of time commences in our singularity, space must emerge, instantaneously producing the physical expansion of space, dependent on the increasing frequency of the wave. We have, in effect, the rapid, dimensional expansion of our singularity that today we label "inflation". The commencing and increasing rate of time created inflation and, with it, the whole volume of the universe, in an instant. Scientists talk about the time it took for this to happen in terms of micro seconds, but the very idea of duration in this period is quite inappropriate. Time throughout inflation ranges from a zero rate in the very beginning to almost today's rate of 1.855×10^{43} Planck times per second, (today's second). So, you can see that any assessment of the time it took for the inflationary process to complete is meaningless from our perspective, in our present day, steady, fast time wave.

I have also proposed that the slight, continuing "expansion" of the universe, defined by the cosmological constant, Lambda is, in reality, the residual quickening of time from this inflationary period immediately after the Big Bang. Everything does take time though (once time passes), so there *was* some overall duration associated with this inflationary period. It is just that we will never "grasp" it from our current perspective, within our wave, with its current rate of temporal progress.

Once this field of energy evolves and forms the continuum, then all the other possibilities that "existed" before the passage of time, now become very unlikely indeed. Since time is now passing, an unlikely occurrence will now take a very long time for it to occur, due to its low probability. Probability is also an aspect of time. With no time, everything is possible and instantaneous, but with the passing of time, only the most likely events happen quickly, the very unlikely ones will take an "eternity". It is this effect that demonstrates how the passage of time "freezes" our universe into being the way it is, from all the possible ways it might have been.

We can see, from the way things developed, that there can be no other universes. This possibility is an illusion in our minds, gained only from the way the universe is, today. We envision space beyond our space, where other universes may have come into being, but space beyond ours does not exist, even if our universe is finite and still expanding.

Beyond our universe there is nothing, the singularity of zero volume and zero time, the nothing that "was" before our singularity erupted with time. This is not a place or a time. It is nothingness. We can also see that a new universe erupting in the middle of ours is also not possible. Our field of time is the creator from which all else emerges and if anything else did have the temerity to try and emerge within this field, it would also be a binary energy field and so it would immediately be assimilated into our field, into our own universe. But the probability of a binary time wave emerging within our time wave is so low as to be equal to zero. Only with zero time will that be inevitable, with a probability of one.

Our wave form of time gives us the classical laws of nature, at the macro scale, above the scale of the wave, i.e. for events slower than the wave. The quantum scale, the scale below the wave form of time, is a realm which more closely resembles the initial singularity, where time did not pass. This is because time does not quite pass here, in accordance with the wave, since events are now shorter than one wavelength, quicker than the wave. So time does not quite pass in the same way it does for us larger, slower beings. No wonder, then, that we have difficulty understanding events in the quantum world. The scale of the wave is the dividing line, the "curtain" between the classical and the quantum realms. The classical realm, with its direct, causal nature is emergent from the quantum world with its probabilistic nature. The background to our universe, to our field of time, must still be there as a type of singularity with no time and no space, existing at the quantum level, below the scale of the wave. This must be the reason why certain actions can occur "outside of time" and which seem to us to be instantaneous and independent of distance. (I refer to the reactions between entangled particles).

During the formation of our energy field, during inflation, a number of possibilities will have "condensed", regarding the different arrangements of these energy quanta and how they can form stable series' of "events" (particles) in space, within and from our wavelike field of energy. With the evolution of the field, this number of possible arrangements is now finite, since the wave will allow only a limited number of ways to curl up, taking energy from the field. Now, the number and types of fundamental particles are limited by the operating energy field within the void. Every event is controlled by the energy wave, which has slowed everything down to a manageable rate and reduced the once infinite number of chaotic possibilities, down to just a few fundamental ones.

The indications are that our wavelike field, the field of time, is the Higgs field. This field permeates everywhere, throughout everything and there is no place to hide from the Higgs. The Higgs field is also reputed to slow the universe down to a sensible pace and so prevents utter chaos. It also gives mass (or inertia) to matter. These phenomena are all aspects that apply to our wavelike time field and this strongly suggests these two fields are one and the same. It would be interesting to work out the value of the Higgs at the event horizon, or at the speed of light. If this idea is correct, then I predict these values would both be zero, since time does not pass under these circumstances, so there is effectively no field locally.

These different ways that energy quanta can arrange themselves, may each correspond to the different "curled up" dimensions in string theory. This number of extra dimensions also matches the number of particles in the standard model. This is no coincidence. Clearly, these extra dimensions are the limited number of possible ways that time and therefore space, can take effect at the very small scale to produce all the standard particles. Remember, everything is made of energy, and if we are right in our speculations, the only place where we can get energy from to make particles, is our wavelike field. There is nowhere else for it to come from.

We deduce that the quarks, photons and other fundamental particles which form the protons, neutrons and electrons (making up all the matter in the early universe), must have "condensed" from our field of energy, during the inflationary period. We also deduce that gravity did not come into effect, until these particles came into existence and that the early part of inflation was uncontrolled, unrestricted by any gravitational attraction. When all these particles had evolved, then universal gravitation had indeed taken effect and the inflation was abruptly halted (almost).

I imagine that, as our wavelike field created space and the field itself expanded physically *with* this inflating space, the wave stretched and broke or otherwise gave up energy, just like breaking waves near the shoreline of a sea. When water waves break, it is because the energy of their previous circular rotation, over deep water, becomes more elliptical as they approach shallow water. Eventually, the structure of the waves can no longer hold together as a whole, and the waves break. Our wave of time will have undergone a similar, if only an analogous, distortion as space rapidly inflated with the wave, so some of the quanta no longer held together. These quanta broke away from the wave form to form particles, the "spray" from the inflating and breaking wave of time. We should remember, however, that the net energy throughout this evolutionary process was always zero, due to the binary nature of the field *and* its emerging contents. The net energy of the *field* is zero and the net energy of every *particle* is also zero, due to their binary nature, for as long as the field lasts.

Only during the initial inflation, was the emergence of matter particles possible. Today, we see the simmering remnants, from this energetic process, as virtual particle pairs, emerging, then quickly disappearing, in the vacuum. These virtual particles cannot endure within a static energy field, where such possibilities are restricted by their low probability. They could only emerge from the inflating field, before the wavelike field enforced its limiting probability onto all possible events. The final, inflated field had done its job and the process of "creation" was now at an end. All events from this point on and all the rules of nature were now cast, irrevocably, by our binary, wavelike field of energy, by the field of time, I suspect, the Higgs field.

With these fundamental particles, continuously drawing their internal energy from the field, we have a reduced local field energy in both binary parts, each particle having its binary nature, with its plus and minus energies, exactly equal to the reduction of the field energies in the positive and negative waves.

The sum total, zero energy of the universe has, therefore, not been altered by the evolution of matter from the field of time. No net creation has occurred, since this law against creation is the overriding law of nature. It is what has controlled the creation of our universe and ultimately defined the very laws of nature.

It is at this point that these ideas join those of the mathematicians and particle physicists, who have worked their way backwards, to this same situation, in a less speculative, more scientific way, from analysing the experimental results from collisions in particle accelerators. I understand, from them, that these particles next merged to form the hydrogen atoms and some helium atoms (filling the primordial void with all the photons that had formed independently). I have proposed that the internal momentum/energy of particles must be fed from the field of time, so it reduces the field's energy, locally to each particle. This reduction in the energy field we know as time dilation, or more precisely, as gravitational time dilation. These hydrogen and helium atoms were the seeds of gravity, made from the energy removed from the field, of the time dilation associated with all matter, and the rest, was inevitable.

The distribution of matter particles, in the early universe, was not completely uniform and why should it have been? Nothing is ever perfect, because everything is fundamentally probabilistic.

All that was necessary, to create the galaxies and star systems we see today, was this slightly uneven spread of matter throughout the volume of space, and gravity did the rest.

Apparently, all the rules needed to create a whole universe are:

1. *There are no rules.* (Anything and everything does happen),

2. *You cannot create.* (You can only create an equal positive with its corresponding mirrored negative),

and

3. *Nothing created is ever quite perfect.* (This stems from the fact that events during the initial stages were based on probabilities, so there are always some natural variations from the ideal mathematical probability distributions).

These ideas and deductions do rely on the universe being a binary universe. If I am right, about the very beginning, then physicists will need to understand and accept the B.U.T. and all its implications, before they can ever achieve a complete understanding, or at least a visualisation, of the very beginning.

Whether these ideas are testable, I am not sure, so we may always be forced to rely on the best speculations, in addition to mathematical proofs and experimental results.

The Big Bang, as I see it, was not some cosmic explosion but more a "Big Fizz". As time commenced, inflation resulted and particles formed from the field, like bubbles on opening a champagne bottle. This is just a simple, layman's, way of trying to understand the process. Certainly, it was a very high energy situation, relative to today, an outward inflation/expansion similar to an explosion. But, this inflation was not driven by any explosive force, or even by some negative gravitational effect, as some have proposed. Inflation was driven by the rapidly increasing rate of time, the increasing of the frequency of the energy wave, from an initial, flat-lined, *red shifted,* non-wave, to the high frequency "pilot wave" of today. It seems to me, to be better described as a sort of "temporal eruption" and associated physical inflation, whilst the basic building blocks of matter "fizzed" into existence, during the early stages of the process.

But What Is All This "Energy" Stuff, Anyway?

I have proposed that the field of time is a field of energy, the source of all energy, so it would be remiss of me if I did not attempt to define exactly *what* this "energy" is. This is a deeply philosophical question which is not easy to answer. In fact, no one has ever answered it, with a fundamental definition of energy. Ask any Physicist or Engineer (or anyone who has studied the physical laws of nature) to define energy and they will tell you that

"Energy is the capacity to do work".

But they are only repeating what they have been told. This statement is not a description of the physical nature of energy, certainly not at the fundamental level. It is a statement of what energy can *do*, its *potential*, and that is a different thing altogether (maybe).

We talk of *potential energy*, the energy a body has, when it is elevated in a gravitational field. We say that it has *potential energy* because, when we let it fall, it gains kinetic energy as it falls and as it loses its potential energy. But the object at height does not possess energy, in the real sense. The potential energy we are talking about is the energy of the gravitational field, which is what does the accelerating, not any positional property of the object, although the two are, of course, related. In any case, the "potential energy" an object might possess is purely relative in nature. If we compare its *potential energy* from a lower elevation,

100 metres down, against its energy from 200 metres down, we find it has twice the *potential energy,* in the second case, than it has in the first. We can see, then that potential energy is purely relative, dependent upon the frame of reference from which it is measured.

Kinetic energy is also, purely, relative. A satellite in orbit could easily be captured by the space shuttle (now withdrawn from service), because they both travelled at the same speed and their relative kinetic energies were both zero. Only an object in front of the satellite, moving much slower would be destroyed by the relative kinetic energy of the satellite. So, we can see, again, that kinetic energy is also purely relative.

Heat energy is just another form of kinetic energy, since the temperature of a substance is a direct measure of the vibration energy of its molecules, their energies of motion, their kinetic energies. If we exclude *potential energy* from our deliberations, because it is never in operation, until it transforms into kinetic energy, then we must conclude that there is only the one type of energy and that is, kinetic energy. Everything else is just another form of kinetic energy. Pressure energy is just the physical result of vibrating molecules impacting the walls of the container and is a measure of their velocity of impact. This is closely linked with temperature under Boyles Law and Charles Law. So, to analyse the nature of energy, we need only to analyse the nature of kinetic energy.

I have shown that the energy used in moving through time is the same energy as is used for moving through space and that as you increase your kinetic energy, then your time energy, your rate of temporal progress, reduces in accordance with the Lorentz transformation. At the speed of light, all the energy of time is being used for speed at the same rate as it is provided by the field and time has stopped for the moving entity. So, the field is capable of doing work and this "work" consists of a combination of temporal progression, a change in state, and progression through space, a change in position, again a change in state. In both cases, whether movement is through time or space, the energy of the field executes work to produce a change in state.

Since the field is continuous, the total energy of the field has to be used for something all the time, otherwise energy would build up somehow and we know that does not happen in the vacuum (except for the vacuum energy, the one Planck time "out of phase" per cycle of the wave). So, if energy from the wave is not being used for physical "change", then it will default into its being used for the progression of time because its energy has to change the state of something, somehow. Irrespective of the degree to which the energy of the wave is being used between these two options, the sum total of both its uses will always add up to the energy of the initial, unreduced value of the wave, the energy of the wave of "universal time".

But even this understanding does not give us a fundamental definition of energy and so we have to go deeper into the rabbit hole to find it. When we look at each individual time quantum, we must ask, "What is it *made* of?" We cannot imagine what it is and so we have to try and eliminate what it is not. It is not a matter particle since matter was formed from the field after the field emerged. It is not a quantum of electricity since electrons also evolved from the field. It cannot be a physical "thing" despite my original assertion that everything in nature has to be a physical entity or process. I now have to concede that the physicality of things is only evident above the scale of the wave since all physical things have evolved from the wave. This is new territory for me, even as I write.

At the beginning of this chapter I made two statements. The first was that "quantum fluctuations" will occur in the vacuum and the second, that they only ever occur in pairs of opposite "sense" in order to avoid a net creation. From this and from the above eliminations, we can deduce that a quantum fluctuation is a "virtual" fluctuation of something, anything, which must always go hand in hand with its equal and opposite, virtual "fluctuation". Quantum fluctuations come in pairs which are purely relative in nature. But, a fluctuation of what? It is not a physical entity, we have already established that. The only thing I can think of which is not physical but yet is *somehow* real, is a "possibility" or a probability. But probability is just the quantification of possibility. It tells us *how* possible a possibility is and so the fundamental is "possibility". It seems that at the deepest quantum level, our reality is composed of mere possibilities which may not have even happened. They are possibilities or potential to either move a system through time by up to one Planck time, or to move it through space by up to one Planck length (at the speed of light).

The Planck time is fundamentally probabilistic.

Time quanta are merely "potential" units of energy which *would* be capable of doing work *if* they existed. This does sound completely daft to me as an Engineer being normally averse to "bull shit", yet I am driven to this conclusion by deductive logic. It simply *has* to be this way since there is no other possible explanation. Mind you, this idea does exactly fit the definition of energy as the "potential" (or capability) to do work and so, in fact, this is precisely what energy is. At the most fundamental level, it is pure potential, nothing more. It seems the universe exists simply because it potentially *can* exist. Well, I never thought I would be saying these things, but I suppose the idea is no more daft than saying "everything can happen and so everything does", that the universe exists because of a version of Murphy's Law[62].

The transition from what is possible to what actually *is*, comes about from the field, above the scale of the field. Below this scale, for events faster than the wave, reality is a little confused from our perspective. Above the scale of the field the classical rules of nature are well defined and our reality emerges from the field of time. Below the scale of the field, each quantum fluctuation is always "trying" to be matched by its opposite fluctuation, but they are never the same at any instant. As each quantum cycles from zero to one Planck time, its opposite is cycling between one Planck time and zero with both averaging one half Planck time at the large, slow scale. It appears the universe, in its attempts to equalise the positive quanta with the negative quanta is "chasing its own tail" and that this perpetual, yet futile attempt, is the driver of time and of change. It is the reason the wave perpetuates, the reason for our continuum. Certainly, at the large, slow scale the waves are indeed equal and opposite, but at the quantum scale, they never are. The quanta are locked in an eternal struggle for equalisation and it seems that the Planck time is the maximum inequality allowed.

One might ask what sets this maximum allowed duration, the "size" of the Planck time, but this question is nonsensical. Such

---

[62] Murphy's Law – A colloquial expression defined as "If anything *can* go wrong, it *will*"

an enquiry has things the wrong way around. We must understand that the Planck time sets the scale for everything. It is the starting point for all events and so it would not matter how "big" the unit is, but only its relative "size" against everything else in our universe. It sets the size of the universe and everything in it. Remember that initially, when the quantum fluctuations were occurring in the void, before our temporal eruption took place, there *were* no rules and that included any "size" of the fluctuations themselves. "Size" is also purely relative and there was nothing to compare the "size" of these quantum fluctuations with.

So the energy quanta that make up the wave of energy, our wave of time, are merely virtual energy quanta or virtual quantum fluctuations and each positive quantum is entangled with its opposite, negative quantum for as long as the wave lasts. The quanta that make up the wave are virtual but in their totality, the wave itself is the driver of time and of events, of change and evolution, of *reality*. The wave of time allows these events or changes, it feeds them, but at the expense of its own energy content, of its own, vast, yet finite potential.

So the standard definition of energy is actually correct and the wavelike field of time really is the "capacity to do work", but I suggest we have now put a deeper meaning to the concept of energy at the most fundamental level.

13 Everyone wants to solve the double slit experiment

The double slit experiment captures everyone's imagination. It challenges our expectations, intuition, and logic and even seems to undermine our ideas about reality itself. Naturally, everyone wants to have a go at explaining the results from a logical viewpoint although no one has yet succeeded. This is indeed a challenge, perhaps the major challenge of today since it encapsulates all the important issues we have with quantum mechanics.

I do not purport to be more intelligent than the best minds in science who have so far failed to propose a logical, intuitive explanation for the results of this experiment. However, my position is presently somewhat unique in that I have an understanding of how our binary universe works and this will surely direct my thinking differently to anyone who has previously made the attempt. Because of this different viewpoint I may be able to make a little more progress than some. At the very least, I may be able to identify different ideas than have so far been proposed.

Certainly the wave nature of both light and particles as indicated by this experiment, indeed the wave behaviour of everything in our universe, strongly indicates that the fundamental of the universe must be a wave of some description and that the environment in which we exist is somehow wavelike. The only plausible candidate for enforcing a wave nature on everything is unquestionably the progress of time, since everything is immersed in the phenomenon and driven by its effects. This logic alone, apart from all the other evidence previously presented, provides a strong case that time is wavelike and I am not aware of any alternative proposal for the cause of this universal wave effect. The closest anyone has come to drawing this conclusion is in Pilot Wave Theory[63], where a pilot wave of some kind is responsible for the behaviour, but no one has yet realised that this pilot wave is the wave of time. I believe that pilot wave theory is the one, correct interpretation of quantum mechanics and I predict that the Copenhagen interpretation will eventually have to develop to match it. Even as I write, pilot wave theory is making something of a comeback and the death knell is already sounding for 'Copenhagen and for some of the daft ideas that go with it.

This one and only available causality for the wave nature of everything which shows up consistently throughout our mathematical workings, results from the wave nature of time and the deductive logic for this is previously described in detail in Chapter ten. The wave nature of time is so blindingly obvious to me that I am somewhat confused as to why no one has yet made the connection.

I shall now use the B.U.T. to try and analyse the double slit results to see if they can be better explained by a wavelike binary time field. If this approach yields the observed results, then this will add yet more weight to the theory. Surely then, this weight of evidence will be enough to convince the most ardent, mainstream physicist to at least *consider* the B.U.T. as a contender.

I do not intend to describe the experiment and its results in minute detail here as this information is freely available on the internet for all to see. I particularly like the video[64] by Prof. Jim Al-Khalili. I recommend you watch it on YouTube before you continue reading. It is only nine minutes long but presents the issues very clearly and with some humour.

To give a general description, monochromatic light is shone through two narrow, parallel slits in a thin plate and onto a screen behind. Behind the plate the two slits are now separate sources of light which then interfere with one another as they radiate at all angles from their exits, spreading outwards and overlapping between the plate and the screen.

---

[63] Pilot Wave Theory – 'otherwise known as the De Broglie-Bohm theory, or Bohmian Mechanics, named after the creators of the theory, Luis de Broglie and David Bohm.

[64] YouTube video – https://www.youtube.com/watch?v=A9tKncAdlHQ

This interference occurs because light is wavelike with peaks and troughs of light intensity as it progresses forward through space. The peaks from one slit (source) are said to cancel with the troughs from the other and this produces the light and dark bands on the screen, the interference pattern. This idea of the destructive interference of light relies on a mistaken belief, a fundamental error in our thinking, which has been completely ignored by all physicists, including Prof. Al-Khalili. The issue has to do with the interference of light with itself.

You cannot destroy or neutralise a photon in the peak of the wave in one beam, just because there are *no* photons in the trough of the wave in the other beam, where the beams intersect. You cannot eliminate a real particle by superimposing nothing onto it and no one it seems has even raised this question about *how* a trough in one wave can cancel or interfere with a peak in the other. This type of "destructive" interference of light is impossible since light is fundamentally particulate. You either have a lot of photons or fewer photons or none at all at any point in the beam. In this regard the wave nature of light is different to waves in a physical medium which *can* interfere both destructively and constructively. Of course "constructive" interference can occur with light when a peak meets a peak and the amplitude (number of photons) is doubled. These are the light bands on the screen. But when a trough meets a trough then nothing is there to interfere as there are still zero photons in both fields at that point and so nothing changes. These are the dark bands we see on the screen. Thirdly, when a peak meets a trough we cannot assume they negate each other like water waves. There are still the photons in the peak of one wave which are not destroyed. These are the mid points between the light and dark bands. Plainly, we have to better understand this issue before we have any chance of understanding the double slit results.

You will see from Figure 26 that our photons exist in the bright bands near the time wave peaks in our binary half, the zone where there is the maximum number of positive time quanta and the greatest positive time energy. But the anti-photons, the photons in the negative time field, cannot exist in these light bands in the troughs of the negative time wave. This is the zone or period where the negative time quanta are at their rarest and where negative time energy is at a minimum. For the negative world to exhibit the same phenomenon as our positive world (i.e. a wavelike light being emitted from its source), the peaks of anti-photons have to reside in the time zone where there is the maximum number of negative time quanta, the greatest negative time energy, i.e. in the dark bands.

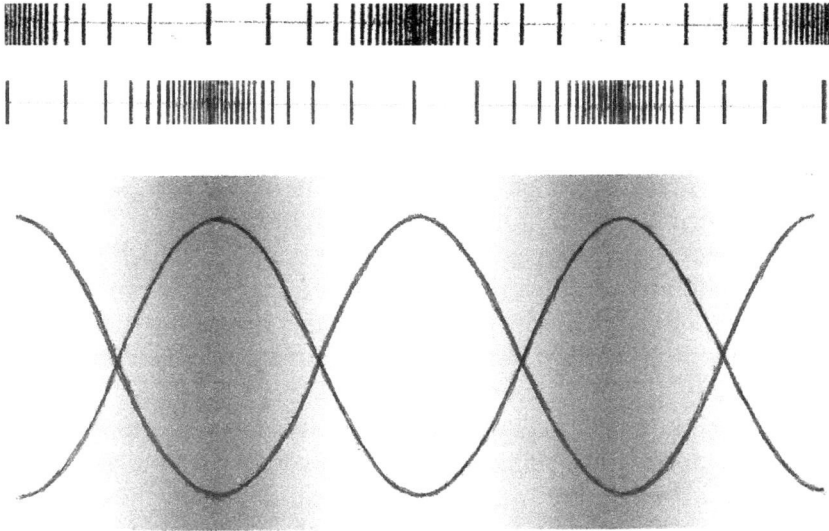

Figure 26: Light waves

Photons are not like other fundamental particles. They are emitted from light sources in waves of varying numbers and we have already established that the greatest number of photons is emitted during the maximum (positive) time rate. The same must also be true in the negative "world" and the greatest number of anti-photons must be emitted during the maximum negative time rate. The zone where the negative time quanta are at their peak is in the dark bands and so this is where the greatest number of anti-photons must reside in the negative world, despite the fact that we cannot normally see them.

So, for the combined field of photons and anti-photons, (i.e. if we were to view photons and anti-photons together as if they were the same thing), then the combined field would be plain and smooth with an even spread of the number of these light quanta.

So, if it were possible to run the two binary time waves together somehow, to negate the one half Planck time "out of phase" between the two waves, then we would see both the positive and the negative worlds, the photons and the anti-photons. In this case, there would be no wave effect of time and no wave effect of photons. In fact, there would no wave effect of anything. There would be an even spread of particles and no wave effect of light. In this instance we might say that the wave function had "collapsed".

It is important to remember that the summation of the positive and negative time waves is also a flat zero all along the axis of duration and time itself loses its wavelike nature in the event that the binary waves run together. Both waves add up to a constant value of zero at all instances. The distribution or number of photons and anti-photons is a direct reflection of the positive wave for photons and the negative wave for anti-photons. There is no wave effect of time or light with the combined fields, but only a wave for each binary part, for positive time separately and for "negative" time, separately.

The anti-photon is exactly the same, physically, as the photon, except, we do not normally see it, since it runs in the negative time wave, in between our positive time quanta. With the photon field, we see a wavelike field (with peaks and troughs of numbers of photons) which is driven by the wavelike field of time. But, in the case where space time has both time waves *exactly* aligned, (i.e. when the one half Planck time "out of phase" is not there), then the combined binary field of time is plain and smooth with no wave, a constant net zero energy throughout duration. In such a case, the photon field will cease to exhibit a wave nature. The space time "interference pattern" is potentially there, *only because* there are two binary halves to our universe and the mathematics of quantum mechanics does show this to be the case. Light has its wavelike properties, only *because* of the wave of time within space time.

Now, the Schrodinger equation states:

$$\hat{H}\Psi(r,t) = i\hbar\frac{\partial}{\partial t}\Psi(r,t)$$

Where *r and t* are values of position and time as the wave function varies over each cycle. The equation means that Ψ, the wave function, multiplied by $\hat{H}$ the "Hamiltonian" operator (a factor representative of the total energy of the wave function), equals, i, the square root of minus one, times Planck's reduced constant $\hbar$ (equals $\frac{h}{2\pi}$), times the rate of change of the negative wave function with respect to time. Before we go any further, we first need to understand what the term "i" really means.

The Square Root of Minus One

The square root of minus one, $\sqrt{-1}$ or "i", is an abstraction we have invented to deal, mathematically, with complex numbers that have a real component and an "imaginary" component.

There can be no such thing as $\sqrt{-1}$ in our positive half of the binary since, +1 squared is +1, and –1 squared is also still +1 and we are stuck, unable to find the answer to the question, "What is the square root of minus one?". We have therefore invented the "imaginary" concept of $\sqrt{-1}$ and called it "i", so that we never have to identify what it is, thus avoiding the problem. This is a very useful tool in mathematics and helps us find solutions to many problems, but the solutions are always what we term "complex". There are always two parts to the answers, one real and the other imaginary. By "imaginary", we do not really know what we mean and mathematicians still struggle to explain this in logical and meaningful terms. The idea of complex numbers first evolved in the sixteenth century. Since then, mathematical techniques have been refined to become useful in many areas of science and engineering. These "complex" numbers crop up all the time in mathematics and it seems to me that the universe has been trying to tell us something through its "language" of mathematics.

With our new knowledge of the binary universe it becomes obvious that the imaginary part of the solution describes phenomena within the negative wave. This part of the solution could be said to be "unreal" or "imaginary" to us "occupants" of the positive wave.

Real terms describing behaviour in our positive wave are simply the real part of the complex number, the terms without "i".

With this in mind, the Schrodinger equation seems to be telling us that the total energy of any system, at any moment (as it varies in a wavelike manner), is defined by the rate of change of the imaginary wave function, at that moment. In short, the energy of everything is cyclical and there are two waves of opposite energy.

Schrodinger is telling us we live in a binary universe!

Let us square both sides of the equation to get rid of the "*i*" term and to see what it means for the combined universe:

$$\left[\hat{H}\Psi(r,t) \right]^2 = -\left[\hbar\frac{\partial}{\partial t}\Psi(r,t) \right]^2$$

This is telling us that the energy of our positive wave squared (at any instant) is proportional to, but negative to, the squared rate of change of the negative wave (at that instant). We might even say that the *probability* of a certain energy level equals the *probability* of the rate of change of energy in the negative universe.

This equality between "energy" and "change" is applicable, not only for any one instant, but also for all instants. So, it is also telling us that the energy of our positive wave over duration (say, one cycle), squared, equals the summation of all the rates of change of the negative wave (over each cycle), squared. The energy of the positive wave is the summation of all the changes over the negative wave and vice versa. This is confirmation that time *is* change. This exactly matches our double wave of energy (time), so it confirms that the energy of time is split two ways, one positive, the other, relatively negative.

Schrodinger tells us:

Time = change

But,   *Time = energy*

Therefore.   *Change = energy*

In the B.U.T., there are always two solutions, a real one from the positive wave and an "imaginary" one from the negative wave.

I find this equivalence with Schrodinger pleasing, but not surprising. After all, I have proposed that time is a binary wave of energy, the pilot wave in question for any system, and that this energy is imparted to the system by the wave form of time as it varies over each cycle. I have said the exact same thing as Schrodinger, from a logical deductive process, as Schrodinger has said in his equation, having worked it out mathematically. We have arrived at the same result coming at the problem from different directions, so, yet again, the B.U.T. is further validated, this time by quantum mechanics.

The very idea of the square root of a negative quantity only makes sense in the context of the square root of something in the binary wave. It makes sense, because a quantity in the negative wave is real, not imaginary, but only when viewed from within the negative wave itself. The negative wave is just as real as ours. It is only imaginary from our perspective in the positive wave, even our mathematics recognises this reality of both the "real" and "imaginary" solutions. Taken separately, viewed from outside of time, both waves are just as real as each other. They both contribute exactly the same energy to all processes, including the progression of time itself. In mathematics, the real part of the solutions to complex numbers expresses relationships within our positive wave and from this perspective only. Imaginary parts of solutions to complex numbers express the relationships in the other, negative, wave. We are the ones who would seem imaginary from *its* perspective. So, I think we now have a clearer understanding of the meaning of the Schrodinger equation, in view of the B.U.T.

Behind the plate with two slits open

Time passes everywhere, of course, and, under normal circumstances, the positive and negative waves are separated by one half Planck time (on average). The wavelike field of energy exists behind our plate, in the experiment, so we should expect light to remain wavelike, as it progresses through space toward the screen. So, we now have two sources of light radiating in all directions, from each slit, in a wavelike manner with "shells" of high photon density interspersed with shells of low photon density. The "interference" occurs, exactly, on impact with the screen, but only because we have conspired to arrange the geometry of the experiment to do this. The double slit experiment only works for narrow slits, below a certain width, spaced apart by a certain distance and at a certain distance away from the screen behind. Change any one of these geometric parameters and the interference pattern does not appear.

We have contrived to make the peaks of the light waves coincide and produce the bright bands in the "interference pattern". Where the troughs of the waves coincide, they produce the dark bands and where a peak coincides with a trough in the opposite wave, these are the mid points between the light and dark bands.

Wave function "collapse" for light

It is speculated that, somehow, a photon might go through both slits. So, to test this idea, the experiment goes on to fire individual particles (electrons) at the slits and to place a detector looking at one slit, in order to observe which slit any particle goes through. The result is clear. The detector "bleeps" fifty percent of the time, demonstrating that particles go through either one slit, or the other, but never both. The important thing is that the interference pattern disappears when we operate the detector, even though both slits are still passing particles. It is said that the presence of the detector (and therefore the very fact that we are observing) "collapses" the wave function. The mainstream speculates about which path each particle takes and concludes it cannot take any path, until the result is known. Physicists are stuck, unable to make sense of the results and, as a consequence of their inability to understand, they *choose to believe* that the path any particle takes cannot be defined until the moment when it hits the screen. This is called the "Copenhagen interpretation" of QM. They see the photons as mere probabilities and not real particles until they actually hit the screen and they then become particles, at that moment. I find this idea to be a serious test of my logic. It seems, to me, to be a sophisticated way of avoiding having to identify the causality. I might accept that we cannot *know* the path each particle takes, until it hits the screen, but to say that, because it cannot be known, is tantamount to the path not being determined, is ludicrous! There has to be a physical, realistic explanation and like Albert Einstein, I prefer to believe the "Moon is still there, even when I am not looking at it". This is my preconception, but you will see I end up agreeing with Copenhagen and that the followers of Copenhagen also end up agreeing with Einstein. They are both right!

Since there is no known mechanism which might affect the paths of the photons, then their paths cannot be affected directly by the presence of the detector, or by any other method of observation. I conclude that it must be the environment, between the plate and the screen, that changes in some way and, in turn, changes the behaviour of the photons. The wave function "collapsing" means that there is now no wave form of time within this region and this space time volume has become un-polarised. The two waves now run together, cancelling each other, perfectly, over duration. The "Pilot Wave" *has* actually "collapsed", in reality. The only way there can be no wave form is if both binary waves are in operation (for both binary parts) and there is no longer any "out of phase" between the binary time waves. This space time has changed from a binary wave and has become smooth and flat. The one half Planck time average separation between the two waves has now been negated, somehow. Of course, with no wave in operation, then this must mean that time no longer passes in the region between the plate and the screen, at least for the particles passing through it. Local space time has become an effective singularity, with no time and therefore, effectively, no space. So now, there is effectively no distance between the plate and the screen for any "occupant" of, or "passenger" through, this space time region. The region has become like our singularity, before creation. It has no time and no space and it has become a realm of only possibilities (or probabilities). The detector, screen, plate and particle have all become entangled, so there is now no time or space between any of them.

In this instance, the paths of the photons no longer spread outwards in all directions upon exiting the slits, since there is no wavelike time in this region to make this happen (I will clarify this in a moment). Basically, from our outside perspective, they move in a straight line (give or take) from the slits to the screen. Remember, it is the wave form of time that gives light its wave nature. If time no longer passes in a wavelike manner, then light will no longer behave like a wave. This is the only plausible explanation for this wave function collapse, but it is not a collapse of any wavelike property that photons or other particles might have, it is the collapse of the wave of time within this space time region. Ultimately, it is the collapse of the pilot wave that leads to the collapse of the probability wave. The screen will then show a cluster of photon impacts with no pattern, peaking at the point dead ahead of each slit, since the photons no longer spread out at the point of emergence from the slits. They only slightly spread out due to the limited degree of freedom, or limited possibilities for their trajectories, that they now have in the non-wavelike environment.

Apart from the fact that photons are emitted from light sources, they are no different to other particles of matter. They are packets of energy, so we should expect the same behaviour from other types of particles. The only difference is that, with light, we are observing trillions of particle impacts with each moment, but with other particles, like electrons, we have the ability to send each one, separately, through the slits and this is what is done next in the experiment.

Wave function "collapse" for particles other than photons

The "interference" pattern is also evident with particles other than photons, such as electrons. Even when we shoot them at the two open slits, one at a time, the pattern still builds up over time, over many events. This demonstrates that the paths of the electrons are determined on a probabilistic basis, showing the probability of an electron hitting a particular point on the screen. More particles hit the screen in areas of high probability than do so in areas of low probability and this is what causes the light and dark bands, *not* any physical interference. Even so, it is obvious that this probability distribution is undeniably caused by some wave like phenomenon. We still call the pattern, on the screen for electrons, an "interference pattern" but, in reality, it is not. It is a probability distribution which nevertheless must still be caused by the wave form of time.

For this effect to be due, solely, to probability, the physical causality must result from the varying amount of deflection on exit from the slits, which must also be probabilistic. This degree of deflection is a chance occurrence of the value and direction of the time wave, over its wave cycle, at the moment of exit. There is a consequential cyclical (wavelike) variation of the electron's momentum and direction, a random selection of direction, dictated by the wave. Clearly, the degree of this deflection will have a wavelike probability distribution over many random events.

Let us not forget that the curving or changing of direction of the path of any entity in space time is *always* caused by the curvature of space time, I would say, the curvature of *time*. When a particle emerges from the slit, it is not deflected mechanically, but, more, it is released into an environment which itself is wavelike and at a particular, random moment in the environment's cycle. At the micro scale, (the quantum scale) time is "curving" all the time, as the time rate changes over each cycle. As a particle emerges, the time rate might be curving in any direction, slowing or quickening, to varying degrees, depending on the position on the time wave at the moment of emergence. That moment is a random event. Any particle, photon, electron, etc., will be deflected towards a trajectory in the direction of reducing time rate, as in the case of gravitation, but here, on the very small, fast scale, in less than one cycle of time energy. The curvature of time, at this scale, is no longer due to the variation in the frequency over a number of wave cycles. It is actually the degree of curvature of the wave itself, at any point (in time), *within* one wave cycle. Normally, it is by this temporal mechanism that light is forced to radiate outwards in *all* directions in waves, from *any* point source, and the emergence of other types of particle from the slits can be no exception.

So, for normal, polarised space time, the wave form of time beyond the slits means that photons and other particles will be "deflected" by an amount in relation to the time rate on the time wave, at the moment of emergence, depending on whether the time rate is slowing down, or speeding up, at that moment. Clearly, this will result in a wavelike probability distribution of the degree of deflection over a large number of events. Even with isolated particles, this wavelike probability will still show up on the screen, if we include enough events to give statistical meaning. With light, of course, with trillions of photon events, we have more than enough statistical samples, immediately, and the "interference pattern" instantly emerges.

Of course, the experiment shows that the pattern does not appear if we have only one slit open, or if we "observe" the emergence with both slits open. So, we deduce that only two, unobserved or unmeasured slits will support the polarised space time beyond the plate. If we "measure" the emergence, then this volume of space time must become non-polarised. There is no wave form of time, locally, without an unobserved pair of slits and so we do not see any wavelike distribution of photons, or other particles, on the screen if we take a measurement.

I deduce that the chosen belief of the mainstream (and it *is* just an *opinion*), that consciousness affects the behaviour of matter, is not quite correct. It is more that the energy effect of a certain type of measurement entangles the elements involved. It entangles the measurement process, the particles and the space time. It changes the polarisation of the space time between the plate and the screen, which in turn, affects the behaviour of matter or light particles in this region.

The wave is evident when only one half of the binary is allowed to prevail, whilst the other half is blocked, in other words, when space time is polarised to expose only the positive wave and not the negative wave. The natural state between the binary time waves *is* of polarisation and we normally experience only our positive wave. But, when it is not polarised we are subject to both waves which we have seen add to zero energy at all instances, so space time, locally, loses its wavelike properties and the pilot wave disappears.

The temporal difference between the waves in the normal, polarised state, is that they are spaced one half Planck time apart (on average) temporally. Of course, the waves are opposite in sense, but equal in value, this is why they add up to a net zero energy (apart from the tiny "out of phase" asynchronicity). When polarisation is destroyed, the two waves act synchronously. They lose their "out of phase" and run together but, in doing so, the net energy of the field, the local vacuum energy, has become absolutely zero and time stops! This must mean that time not only stops for the photons, due to their inertial time dilation, but it must also stop within the locality of the non-polarised space time, so all types of particle must hit the screen instantaneously. Remember, it is the *wave* of time that drives time progression into the future. With no wave, there is no temporal progression and on "collapse", the wave has completely *red shifted* into a "straight line", a constant value of zero energy over duration.

As mentioned previously, the definition of entanglement is when the entities in an entangled system share the same wave function, there is no time experienced between these entities and their relationships occur outside of time, instantaneously. This is the case here, with the two waves exactly synchronised and with no half Planck time "out of phase".

We assume that photons, at least, travel from the slits to the screen at the speed of light. But, if time has stopped locally, this speed, relative to us, has become infinite, not just for the photon. Other particles, like electrons, normally travel slower but even these, on the collapse of the wave function, now travel instantaneously to the screen. In fact, the idea that their paths are undetermined, until they hit the screen, has now gained some real physical meaning in my mind. It may as well *be* that their paths are indeterminate, since their journey is instantaneous. There simply *are* no paths taken. It may as well *be* that the particle, whatever it is, is merely probabilistic, since it does not actually exist along a path, but only at the slits and at the screen. It is now clear to me that this is what the mathematics of quantum mechanics is saying. The Copenhagen interpretation almost gets it right, but not quite. It is the wave nature of space time that gives us these results, not some wavelike property of the particles travelling through it. As a reminder, for two entangled particles, the distance between them does not matter, nor does the timing of some input to one particle, as the other particle must react instantly since they are linked "outside of time". There is no effective space between them and they are effectively "touching".

In summary so far, according to the B.U.T. our observations are telling us:

- When there are two, unobserved slits emitting particles, the local space time is normal and polarised and time passes in a wavelike manner between the plate and the screen.

- When there is only one slit open, local space time is not polarised and there is no longer any wave nature of time.

- When there are two slits open, but a measurement is made, local space time is also not polarised.

I shall explain the causality for how these situations occur in a moment, although as a passing thought, I have it in mind that the Casimir[65] effect is also brought about by this situation, where the space time between the plates is no longer polarised. When the two plates are in very close proximity, the time wave between them ceases and the energy of the time wave exterior to the plates now exerts a net force, against which there is now no internal energy to resist. There is a resulting, net attraction between the two plates. The mainstream explanation is that there is "*an imbalance of vacuum energy*" between the outside of (and in between) the plates and this is exactly what I am saying. I am asserting that the vacuum energy is due to the polarised space time, with the one half Planck time out of phase in operation. Remove that and you will get a net imbalance. There must be a similar attraction between the plate and screen, in the double slit experiment, when the space time is non-polarised, but this is so small it has gone unnoticed, although this could now be measured to test of this hypothesis. Positive results from this test will verify this analysis of the double slit experiment and will prove the Binary Universe Theory.

In the macro environment, space time is normally polarised. Time runs in a wavelike manner and the wave nature of light is evidence of this. This is the state of the polarised nature of the local space time, when both slits are open, and no measurement is being made. However, when we make a measurement, all the information within the local space time is being linked with the measuring system apparatus and the whole, local environment becomes entangled with it. The energy of the one half Planck time "out of phase" is removed and used for measurement and the wave form of time disappears, We have seen that the definition of entanglement is that all entities, within an entangled "system", share the same wave function, the result being that there is effectively no time and no space between the participants of the system. Effectively, they are all "touching". The local system, including the particles and the space time, have become entangled and the distance between them has become irrelevant. To thoroughly grasp these ideas and to complete the analysis, it is always best to look at things from an energy perspective.

---

[65] Casimir Effect - The attractive force between two closely positioned plates.

From an energy point of view

The easiest way to understand the workings of any system is to look at it from an energy balance point of view. All systems must comply with the conservation of energy and this example can be no exception. So, let us see what else, if anything, such an analysis might reveal.

The energy contained in the space time, in between the plate and the screen, can only be reduced by something removing that energy, in order to use it in some other way. We have seen that all matter, all processes, get their energy from the field of time and there are no exceptions to this. Remember, the net energy of the two binary waves is zero (almost), so there is only a minute difference of one half Planck time in the space time energy content, between a system with the waves running one half Planck time out of phase, and a system where they are running together, precisely. This energy is, in fact, the vacuum energy. The experiment tells us that the energy demanded from the system, by any measuring process, is at least enough to disrupt the waves by one half Planck time, so it removes the vacuum energy. We still comply with the principle of conservation of energy. The very small, almost zero net energy of the unmeasured space time, equals the energy drawn from the space time to power the measurement process, plus any waste energy given off by the system, somehow. The measuring process has become entangled with the rest of the system.

It is important to understand that the measuring system has to remove energy from local space time, or the out of phase will not disappear. This explains why – when we leave the detectors in place, but disconnect the measuring system from its power supply, the wave no longer collapses, since there is no energy demand from the measurement. It has become a non-measurement and so there is now no energy drawn from the space time. It has nothing to do with any conscious observer.

For the situation with only one open slit, again the local space time must be non-polarised. The binary waves must, again, run together. So, in this situation the question is, "Where does the demand come from to extract this energy from the space time?" There is no measurement in progress, after all. The only energy being demanded from the space time, in this state, is from the particles emerging from the slit. So, their energy demand (from the energy of the field of time) must also result in the loss of the one half Planck time, "out of phase" between the time waves. The energy of the particle is taken from the local space time field and so the vacuum energy is destroyed when the particle enters from the slit. In short, the particle becomes entangled with the rest of the system and the pilot wave disappears.

But why does this not happen with both slits open?

With both slits open, the space time runs as normal with the waves out of phase. We can deduce that there is no net energy demanded from the space time in this situation or, at least, that the energy is demanded in complete Planck time units and not just one half Planck times. This must mean that any energy demand from one particle emerging from a slit must be replaced somehow by other particles entering from the other slit. The energy demand from each particle negates the one half Planck time out of phase, but another particle, entering from the other slit, negates *another* one half Planck time and this brings back the out of phase for the waves.

But it is not quite as simple as that. Remember that the energy of each time quantum varies sinusoidally between zero and one Planck time and so this "second" half of a Planck time is not just any old half of a Planck time, it is the other half of the quantum couple and always makes up the fraction of the first time quantum to one Planck time! It is the "anti" of the first fraction, the negative quantum of the positive quantum, the negative quantum entangled with its positive quantum.

This tells us that, for every particle emerging from one slit, its anti-particle enters from the other slit over a large number of events. But how does this perfect match between quanta happen? It certainly cannot be a coincidence of this magnitude, so there must be a cause. Thinking back to the B.U.T.'s interpretation of Schrodinger, we remember that:

> *The probability of any particular energy level*
> *equals the probability of its entangled*
> *negative rate of change in the opposite wave.*

But exactly what does this mean? It means that if you take a large number of random samples of energy levels, there will be the same number of associated, entangled changes in the opposite wave. When particles emerge from a slit, they do so at a particular point on the time wave, at a particular energy level. So, over many such events, there will be the same number of associated, entangled particles emerging from the slits, the same number of anti-particles. The space-time energy is reduced by only complete quantum couples, or complete Planck times and never just by one part of the binary energy, as is the case with only one slit open.

The space time energy is reduced, but always by one complete Planck time and the net effect of this is, always, to maintain the polarised space-time. With two slits emitting particles, the space time polarisation is constantly maintained, simply because there are *two* slits and not just one.

The two waves are reduced, certainly, but both to exactly the same extent and so the space time remains polarised (over time). This balance of energy demand occurs over a large number of pairs of emerging particles. This does not mean that the combined energy demand of one complete Planck time takes place with each consecutive pair of particles, but only that, statistically, this must balance out over a large number of events. So, if we were to compare only two sequential events, we might be lucky and get a complete Planck time energy demand, but we would be just as likely get an imbalance. So, it is purely the particle pairs emerging from the slits, removing only complete Planck times over many events, that results in maintaining the out of phase of the time waves. In this situation, with both slits open, the polarised space time between the slits and the screen is maintained, with a consequential wavelike probability distribution of particle impacts on the screen.

This completes the analysis using the B.U.T. and I have given here a logical, intuitive explanation for all the effects from the double slit experiment *without* recourse to any daft ideas about "many worlds" or particles being in two places at once, or consciousness having a physical effect on events.

I have shown that the Copenhagen interpretation is almost correct, in that there really are no paths taken by the particles and so it may as well *be* that they take all possible paths. I have shown that particles only exist at the slits and at the screen and so it may as well *be* that they only exist as probabilities until they hit the screen. The problem with Copenhagen is that it has no causal explanation for these unreal particles and their probabilistic paths, nor does it explain how the wave function collapses or exactly what the wave in question is. Pilot wave theory does propose a real wave of some sort but again it falls short of defining what this wave is.

I have applied the B.U.T. to the problem and this has predicted the exact same results that we get from the experiment.

This analysis brings together, Einstein's real-world view, the Copenhagen interpretation and Pilot Wave Theory with the double slit results, in a logical way and with a clear, understandable physical causality for every aspect. The understandings of Einstein, Bohr, Heisenberg and Bohm are all correct, (if somewhat incomplete) and we now have a full, causal explanation for the results of this experiment, which brings together all the major interpretations of QM.

All these ideas flow, logically, from a full understanding of the B.U.T., I therefore assert that the Binary Universe Theory is finally, and unquestionably, validated by the resolution of this important, present day conundrum. This not only shows that the B.U.T. is almost certainly true, since it would likely not stand such a test if it were false, but that it also shows the power of the English language in the process of deductive reasoning.

Whilst we rely almost entirely these days, on the mathematical process to show us new ways of thinking, we should never disregard the power of the mind, of language and of the written word. In this case, mathematics has brought us so far in terms of understanding but it cannot take us further. It is our *interpretation* of the math, using the B.U.T., that completes the picture. Language and the human thought process may not be as precise as mathematics but it can have an even greater predictive power when used in conjunction with deductive reasoning, visualisation and imagination, and more especially, when it still aligns with the mathematics.

Scientists are sceptics, of course, and they will no doubt respond by saying that this explanation relies on a binary universe which is not proven science. But scepticism gets you nowhere. It is imagination that produces the ideas for the sceptic to challenge and test. The combination of these, imagination and scepticism, we know as science. Both scepticism *and* imagination are essential for our scientific progress.

These days, the scepticism of the scientific mainstream prohibits the acceptance of new ideas by default. It will pick on one aspect of a new idea which *seems* to contradict accepted theory and immediately throw it out, without entertaining it in the entirety of its context. This is not science. It is not even scepticism. It is prejudice, an unreasonable resistance to changing our thinking, even in the face of undeniable evidence. But, the fear of having to change our way of looking at the world should be dwarfed by the consequences of denying the B.U.T. and having it prevail ultimately. The protagonists of such a view would clearly become the subject of ridicule in the history books of the future. To avoid this, the mainstream's reaction to these ideas must be carefully thought through and the B.U.T. *must* have its impartial hearing for the sake of all.

14 The case for the binary universe

The B.U.T. aligns perfectly with Special and General Relativity. There is no conflict with these two theories. The differences are matters of opinion or interpretation and there are no issues with the mathematical workings.

I have presented how we might interpret the curvature of space in strong gravitational fields as the increasingly rapid changing of time rate with position. This is an alternative view to the traditional geometric interpretation of GR but both interpretations will work equally well. In fact, the special and general theories are completely unaffected by the acceptance of these new interpretations and they still stand as testament to the great mind of Albert Einstein.

I have presented the evidence for the Binary Universe Theory. Any speculation in the latter parts of the book was used only to present how the theory might fit with many aspects of accepted science and to explain certain issues that still perplex us today. The disagreements are matters of opinion, of belief. There is no dispute with the mathematics.

The proposal for a binary, quantum-like time can be checked mathematically against Quantum Mechanical theory as well as with the standard model and its elusive super symmetric partner particles.

Certainly, the main ideas I regard as compelling.

The B.U.T. is, quite frankly, obvious.

The main things preventing acceptance are the current, entrenched beliefs that all motion is purely relative and that there is no preferred reference frame. Again, my suggested experiments to demonstrate the *blue shift* observed from the more accelerated frame and the existence of the time rate field (the preferred reference frame), are the only things that will settle the debate. The B.U.T is testable and I have presented how to test it. I have also removed any fears about the preferred reference frame and the temporal acceleration to the relative, infinite speed in the moving frame, by detailing how our universe still works just the same, in accordance to all the known laws of physics.

The B.U.T requires fewer entities and eliminates the Graviton and Newton's force. Gravity, as a force field, is replaced by the effects of time dilation.

The abstract curvature of space in GR is simply the result of the extreme curvature of time in strong fields. It is, therefore, the simplest theory and so, according to Occam's razor, it is more likely the correct one.

Newton first invented the gravitational force to try and explain gravitational behaviour and he employed the idea of this force "acting at a distance", as the indirect cause of gravitational acceleration.

He had no choice at the time.

He knew what the rules of behaviour were, but could not explain the causality. It was not until the early twentieth century that we became aware of the phenomenon of gravitational time dilation and it is, perhaps, understandable that we have held on to our belief in the gravitational field and simply "tagged" time dilation onto it.

It is human nature to take comfort from the familiar, so we tend to give more credence to those ideas first presented (and long accepted), as opposed to those from newer propositions. We all exhibit some chronological bias in our judgements.

When certain ways of thinking have been established for many years, we sometimes find it difficult to change them, to accept new ways of looking at the same thing.

We must find courage and realise that the gravitational field is a *creation*. We cannot detect it or measure it, but only infer it from observed behaviour and there is actually no scientific proof of its existence as a physical field.

General relativity describes the behaviour within the field. It does not define what the field is. Newton's Law of gravitation is not a law of nature, but is, again, just a rule of behaviour. A law of nature includes the cause and effect but a rule of behaviour does not have to. It can be merely empirical (with arbitrary constants) and yet still describe the behaviour.

The gravitational field has evolved from Newtonian thinking and is a fictitious field. It is not a real field which acts on particles and enforces action but an abstract field that predicts behaviour. Newton's gravitational force is also a *creation*, which does not comply with causality and so cannot be scientifically justified. The curvature of space in GR, stems from the equal treatment given to the four "dimensions" of space-time and so, the idea is suspect, despite the fact that the mathematics works. Space and time curvature is indeed an effective way of describing space time, but we should not believe they are both as fundamental as each other. Only one can be fundamental, whilst the other is an emergent effect and I have demonstrated here that, of the four "dimensions", time is the *only* fundamental.

Evidence of this comes from the one, verifiable phenomenon we *do* know exists in a "gravitational field", the time curvature, which *is* directly measurable and is therefore most definitely *real*. Time curvature is the *only* phenomenon, within a "gravitational field", which can be directly observed, or detected, and I have shown here how this must be the cause for any and all effects on free fall motion, even in strong gravitational fields.

Since the early part of the twentieth century, we have made no attempt to reconcile the time dilation field with our imaginary gravitational field. We have chosen to believe that gravity somehow causes time dilation and to blindly accept that you cannot have one without the other.

We have been unable to demonstrate *why* this is the case, or to establish the connection between these two "fields". Nevertheless, I have shown, here, how these two ideas are reconciled, by the realisation that only the phenomenon of time is real. The curvature of space is the secondary effect from the time curvature. Space curvature is perhaps more than just an abstract interpretation of these effects but is still only an emergent phenomenon from the field of time, with its changing rate across distance.

The new theory proposes that there is no gravitational field at all but only the time rate field and that this is all that is required to produce the observed behaviour of free falling particles in both weak and strong "gravitational fields". I have presented mathematically how this purely temporal view can adjust both Special and General Relativity to give us a better understanding of our temporal world.

The "graviton", yet another creation, was an attempt to prop up the failing notion of the imaginary gravitational force field. It is highly questionable to assume that an effect which stretches to infinity can have the same type of mechanism, a particle, as other effects which operate on much smaller scales and even at the subatomic level. This is more evident from the proposed differences between the nature of time above and below the scale of electromagnetic radiation. That is to say, time is a sinusoidal wave form at the macro scale but becomes increasingly influenced by its quantum-like nature below the scale of EMR and more so, the closer we get to the Planck scale.

At this small/fast scale, events do not "see" the wave, but are controlled by only fragments of it.

There are ongoing debates about materialism versus quantum physics generated by the false notion that we should be able to apply the rules of the quantum world to the macro world. The wave form of time creates the classical rules of nature, but the wave only has full effect above its own scale, for events which take longer than one wave cycle. For events quicker than this and so for the world smaller than this, then the rules of nature cannot be defined by the wave, but only by fragments of it and ultimately by the quantum nature of time. Attempts to apply the rules of QM rigidly to the macro world are clearly inappropriate, as are any attempts to describe the world of the small and fast with the curvature of time (and space) and the classical laws of nature.

The idea of the graviton somehow providing gravitational attraction is inconsistent with, and even in competition with, gravitation being the result of the curvature of space-time. This remains a fundamental conflict in science today, although the graviton does now seem to be mysteriously disappearing from the scientific menu. The graviton will soon be "off". If there is a particle emanating from the field which produces the phenomenon of gravitation, then that particle, according to the B.U.T. will be the Higgs boson emanating from the Higgs field, the field of energy which I believe is the field of time.

Gravitational acceleration is currently understood to be due to the distortion of space and time from General Relativity, with particles (objects) following lines of least action (geodesics or world lines) through space time. Although this idea, developed from the math of relativity does give correct predictions, it is merely one way of looking at free fall motion, which can be conveniently manipulated with mathematics. In reality, there is no such physical entity as a geodesic, no set of rails that directs the motion of free falling bodies. The geodesic is yet another creation, a handy mathematical rule of behaviour emanating from General Relativity.

The initial interpretation of Special Relativity in 1905 immediately took a wrong turn, having been misled by the preceding nineteenth century notion of length contraction. Special Relativity accepts the phenomenon of length contraction due to motion as being as real as anything else, paying no heed to its purely relative nature. We are thus told "everything is relative", in order to support this belief. As a consequence, we have failed to understand the full significance of time dilation despite the clear evidence from the Hafele & Keating experiment which demonstrates that time dilation is a real phenomenon (as well as being relative). Not only have we ignored the obvious reality of time dilation, but we have also actively conspired to diminish its reality in our thinking, in order to cling to the belief in all things relative. We have engaged in wilful, self-deception born of the fear of the unknown.

The mainstream's chosen view of SR is driven by the fear of superluminal speed and of there being a preferred reference frame. Any proposition leading to these ideas is immediately met with an irrational resistance, before even entertaining it, let alone accepting it. When irrefutable evidence is presented to challenge the mainstream view, like the H&K results, the mainstream defends its position with unjustifiable claims like, "Well, SR is counter-intuitive", and we are sent away believing we are not intellectually capable of understanding it, or so the mainstream would have us believe. But they are the frightened ones, not I.

With General Relativity, the final defence against any challenge based on deductive reasoning is to throw the mathematics in the face of the challenger. "Do the math" they say, "Do the math. If you cannot do the math then you cannot argue against it". Well, I can do the math and I am not arguing against the mathematics. Physics is the study of the physical world and however complex the math might be, the world must still be comprehensible using deductive logic and the rules of nature, including the Causality Principle. I just do not "buy" the idea that our intuition cannot be applicable to complex science just because intuition was given to us so that we ran away from lions on the savannah. Common sense deduction can be applied to anything if only we know enough about the realities to apply it. I am not disputing the validity of the mathematics of GR in terms of its ability to make correct predictions. I am suggesting a modification to our understanding of it, which gives the same answers, but which also demonstrates causality and reflects reality at all stages. This has to be an improvement.

I am an Engineer, trained to produce results and to make progress. I am a problem solver with both feet firmly on the ground. I do not take kindly to being told that I cannot expect the universe to be understandable. I will never believe that the universe is incomprehensible. If there is something we do not understand then that is because we do not yet have enough knowledge, or we have a mistaken belief somewhere, not because anything, inherently, defies comprehension.

Some quantum physicists seem to revel in this apparent incomprehensibility, fawning over aspects of theory which defy logic. "What a marvellous theory! We cannot understand it!. Look how counter-intuitive it is". This "Monty Pythonesque[66]" outlook seems perverse to me and I oppose it. I see these "counter-intuitive" aspects of science as challenges to which we must rise, comprehend and explain. I believe that they are understandable and that we have the intellect to master them with a little more time and a few changes in our thinking.

Inevitably, arrogance in Mainstream Physics will lead to confused, irrational thinking which is neither desirable nor justified. Eventually, there will be an error exposed, somewhere.

Can the Mainstream train of thinking really be as follows?

> *"We cannot allow speed in excess of light speed under any circumstances. The speed of light must remain invariant... Therefore, we must decide on length contraction with relative motion and keep the velocities the same in both frames.*

---

[66] Monty Pythonesque – after a popular UK 70's TV comedy *Monty Python's Flying Circus* – extremely surreal interpretations of popular beliefs and/or normal events

If we allowed the alternative acceleration, induced by the time dilation, we would get superluminal speed and ultimately infinite speed in the moving frame and we cannot allow that. The speed of light must remain invariant from all perspectives."

"Because we have opted for length contraction, we must use this to justify experimental observations like the prolonged life of muons and we must ignore the fact that the time dilation explains this completely, without requiring length contraction."

"To protect this approach, we must devalue the phenomenon of time dilation, so no one can take the alternative view and maybe no one will notice we have used two mechanisms to explain the same phenomenon, when only one is required. We will claim that time dilation is no more real than length contraction, despite H&K."

"We must defend this approach against the argument that time dilation is real. We must claim that all things are relative: time dilation, length contraction and motion itself."

"We can therefore claim there is no preferred reference frame by declaring all motion to be purely relative."

"If anyone argues against this view using common sense argument, we will claim that relativity is not necessarily "intuitive". Even our use of the word "intuitive" is a "put down" and will devalue opposing argument making any deductive logic used seem like "intuition", a much less rigorous thought process than deductive reasoning."

This view would clearly be paranoid and I have exaggerated, to emphasise the point. Of course, there is no real conscious conspiracy.

The point is that, over the years, any mainstream belief always self-protects and presents arguments to defend itself against challenges, as they arise. But, if the mainstream view is flawed, then somewhere along the line after piling up a flawed defence, some fatal error will inevitably be exposed.

In this case the error is within the mainstream's interpretation of SR and it was exposed by the 1971 Hafele & Keating experimental results. These indisputable results from H&K show that the moving clocks got there quicker than the time elapsed on the stationary clocks, for the same journey. This must mean there was a relative acceleration in the moving frame caused by the time dilation. The velocity is different in the moving frame, not the same as in the non-accelerated frame, as is understood by the traditional view.

The fundamental mistake in the mainstream understanding of relativity is a philosophical one, it is common to both Special and General Relativity. The error is best understood by considering how observed lengths or distances may as well be shorter if your time is slower in the moving frame.

The key thing here, is that the mainstream admits that proper lengths never change and so how can they claim that any length is ever really contracted!

I argue that only the time dilation is necessary, to explain everything. The geometry does offer us the two options, but length contraction has been given the same status as time dilation and this equality is unrealistic. This *declared* equality does not require superluminal speeds in the moving frame and avoids a *preferred* reference frame.

Preconceptions have been "fed" into the theory, in order to get the desired result and this is not science. This leads to a conflict within Special Relativity (when at the speed of light, when it is accepted that time stops, the speed experienced in the moving frame is not taken to be infinite).

How *can* speed not be infinite if your clock has stopped and you get anywhere in no time at all? SR is clearly wrong in regarding length contraction to be as fundamental as time dilation.

The conflict is removed by understanding:

> *It is only the clock which is directly affected by motion, but lengths are merely affected by the clock.*

This is evident from our new understanding of the time rate field laid out here. The mathematics works well for both ideas, for both the right one, and the wrong one. A similar misinterpretation is duplicated in General Relativity by the claim that space, as well as time, is curved by massive objects, yet no causality is suggested for this space curvature. Again, it is a preconception fed into the mathematics. This gives us correct answers, but it bypasses the logical deductive sequence of discovery, avoiding having to explain this curvature of nothing.

I have presented here the causality for how time is slowed in the vicinity of mass.

The energy of the time rate field is diminished by its partial use for the internal energy of matter particles. Until now, no one had ever presented the causality for the physical curvature of space and I have now presented how the rapidly changing time rate in strong gravitational fields does lead to the effective curvature of space. This new knowledge leads us to the fundamental nature of time and the fact that space is emergent from time.

Lengths are envisioned in GR as being curved but this is purely due to the time dilation within the gravitational field and again, the mathematics works for both ideas. Space itself can therefore be regarded as being more physically "curved" (compressed), the deeper into the field you go and as the "gravitational" time dilation rapidly increases over distance. Strangely, this means that space itself becomes "too big to fit" and so it has to become curved (mathematically anyway), to align with the fact that it takes longer to get from A to B relative to an observer further out in the field.

The absence of a cause for space curvature has persisted for over a century because the mathematics works for the secondary effects of length contraction and its offspring, the curvature of space. But the math also works for the correct view of inertial and gravitational time dilation and it is high time we brought our thinking into line with reality.

If we do not do this, then it will be impossible to create a plausible theory of everything. Our understanding of space and time will remain forever flawed.

It is never too late to change direction, if we see a chance to improve our ideas about reality and the B.U.T. now offers the opportunity to reconsider how time dilation affects inertial and gravitational behaviour and to present the mathematics of General Relativity in terms of the real energy of time instead of the abstract geometry of space.

The very process of matching GR with other areas of proven science and of maintaining mathematical rigour has no doubt led to the accurate predictions which GR provides. It is *as if* space itself is distorted and in such a way as to align with predictions from the general theory.

Nevertheless, just because GR gives the right answers, does not mean that it must reflect reality at all stages of its workings. GR is a predictive rule of behaviour, not a descriptive law of nature and there is a subtle yet important difference between the two. It was a very clever move to invoke a curved space time in the development of the theory and the inbuilt rules of nature coupled with mathematical rigour made sure it could not be "wrong". Nevertheless, I emphasise again:

> *Abstract ideas cannot accelerate physical objects*

and,

> *Only a proximate, antecedent cause can affect motion*

The geometric interpretation from General Relativity does not provide this causality. It only predicts the behaviour!

Newton's law of gravity is another example of a theory which matches observations and gives "accurate" predictions whilst not defining the cause and effect, but at least Newton knew that. Perhaps this example is relatively easy to come to terms with, especially since Newton himself acknowledged it.

In GR, the curvature of space time is a useful idea for making predictions, but it is the extreme curvature of time that gives rise to the effective curvature of space.

Time is the only real operator within space time.

The new theory, presented here, shows that time dilation is the cause of gravitational acceleration and that there is, in reality, no gravitational field, no gravitational force and no need for the graviton.

It is Newtonian gravitation that is real, but it is only real *within* any frame under consideration, (i.e. at any infinitesimal point within the field). Farther away from any event, motion is distorted by the effects of time curvature but this distortion is again due to the same time curvature which creates the field and it is equivalent to the distortion of space described by General Relativity.

For any observations in a gravitational field, we see reality directly if we share the same frame of the event, but observations from a distance are always distorted by the "lens" of time curvature.

The deductive reasoning proposed and used here, to explain the mechanism of gravitation, involves only time curvature and its effect on linear, constant velocity. It is accepted today that Newtonian gravitation (weak fields), can indeed be viewed as time curvature.

In other words, for Newtonian gravitation, the mainstream accepts that changing the relative lengths of the seconds will change the relative velocity and that this is what causes Newtonian gravitational acceleration. This is accepted by all, so we must also accept the same effect on velocity in Special Relativity.

We know that time slows with increasing speed and so the seconds also get longer for inertial systems too.

In this case, we will have the same increase in velocity within the frame of fast moving objects induced by this "curvature" of time, but produced by acceleration rather than by a gravitational field.

This argument directly reflects the principle of equivalence. Since gravitation is also viewed as an acceleration effect (an equivalence with acceleration), then this demonstrates that the accepted view of Special Relativity is inconsistent with General Relativity in holding velocity constant between frames.

You cannot accept acceleration due to time curvature for gravitational fields and yet refuse to accept it for inertial systems. If you accept the one, you have to accept the other!

This being the case, the modifications to relativity theory proposed in this book, all fall out from this one basic premise and ultimately, we are driven to conclude the following:

- Special Relativity is incorrect in its choice of a common velocity in each frame. Velocities cannot be the same if the time rate is different in each frame

- In the moving frame there is no speed limit and velocity tends towards infinite speed, due to time dilation whilst it tends towards the speed of light as viewed from any frame other than the moving one.

- Time dilation is the only available cause for any effect in space-time. Length contraction is a purely relative illusion which can never be a cause for anything.

- Time is quantised at the Planck scale but wave like at the scale of EMR due to the cyclical "compression" and "expansion" of the time quanta.

- The preferred reference frame is the time rate field, the *standing* wave of time.

- Inertial time dilation is caused by movement through this *standing* wave of time, the preferred reference frame.

- The wave nature of light is caused by the wave nature of time. It is actually a mirror of the wave of time.

- The speed of light is not precisely invariant but the rate of universal time is invariant (ignoring the slight speeding up of the time rate causing universal expansion).

- Newtonian gravitation is caused by time dilation and is given by the universal law:

$$g = \frac{c^2}{2r}\left(1 - t^2\right)$$

- *Red shift* or time dilation is present in all gravitational fields because it is the cause of gravitation. In reality, there is no such thing as a gravitational field but only the time curvature, the time rate field, the changing energy of time over distance.

- The curvature of space is an effect from the curvature of time and is only noticeable when the changing of time rate is significant over small distances. It is only time which varies within space time, so time is the single cause for all effects. In a gravitational field, the deviations from Newtonian behaviour are caused by the high rate of time curvature. A temporal understanding of General Relativity is proposed, based on this idea.

- Gravitational time dilation is fundamentally caused by the spin energy, or internal kinetic energy of subatomic particles. Every particle draws this energy from the field of time and causes a minute dilation of time locally. This created the first stars from the primordial void. With many atoms in large masses, these minute effects combine and accumulate in all directions to produce the time dilation at the surface of the mass. The resulting time dilation field above the surface is the proximate, antecedent cause of Newtonian gravitation.

- The time dilation field (gravitational field) is caused by the transfer of time quanta from temporal duties to inertial duties for the momentum within the particles of matter and results in localised gravitational time dilation and time curvature within the field of time.

- Our universe is a digital binary pair with each part existing in between the other's time quanta.

- All energy is fundamentally, time energy.

- The energy of time is quantised and finite (over time).

- This finite energy is the cause of the finite limit to speed "c".

- Time is a real energy field but space is emergent from time. The energy of time is the fundamental for everything.

- The field of time is wavelike and this is what gives particles their wave nature.

- The wave nature of time is deduced from the wave form of light.

The B.U.T. challenges some entrenched, conventional beliefs in relativity and I have no doubt it will provoke objections.

Some may be afraid to align themselves with the notion of superluminal speed, even though this occurs only in the moving frame and is purely due to time dilation.

Others may find the idea of a preferred reference frame hard to accept having born witness to the protracted, historical deliberations to "finally" rid ourselves of the Aether. Maybe some will find it impossible to accept that all motion is not purely relative after all and they may have difficulty in visualising the *standing* wave of time. Certainly, the predicted *blue shift* of the stationary frame will provoke objections. Protagonists of General Relativity may feel protective towards their beloved theory and towards the reputation of Albert Einstein, but it must be remembered that this new theory in no way diminishes the validity of GR nor the stature of Einstein's great achievements.

If GR had not been created then most of the technological progress in the last one hundred years may not have happened and GR has provided us with one huge, practical step in our understanding of the world. But, it is now time for the next step, time to move on. If anyone has any difficulty at all in accepting these proposals, then they should ask themselves one simple question:

> *"Do I really expect to be flattened in the fore/aft direction as I approach light-speed?"*

I am certain the honest answer to this will be "No", since the laws of nature must remain intact at *any* speed. You must accept though, that your watch will have lost time when you arrive home. The point is that length contraction is an illusion, but the time dilation is *real* and everything else in the new theory falls out naturally from this one, subtly different understanding in relativity.

The mistaken belief in real length contraction is what has been holding us up for over one hundred years and once we get rid of it, all the ideas in this book evolve before our eyes.

The final nail in the coffin for the mistaken beliefs of the mainstream is the complete compliance with causality in the B.U.T., something even General Relativity cannot lay claim to. Yet I have no doubt that that the debate will continue until the proposed experiments are carried out. This is despite the B.U.T. giving causal explanations for the following present day problems with our understanding of nature:

- The stability of the continuum
- The wave nature of light
- The antimatter imbalance
- CPT Symmetry
- Quantum entanglement
- Something from nothing
- Inertial time dilation
- The preferred reference field
- The vacuum catastrophe
- The cosmological constant
- The expansion of cosmic space (Dark Energy)
- The curvature of space and time
- The variation in speed of different frequencies of EMR
- Invariance of the speed of time
- The equivalence principle
- The graviton does not exist
- What is inertia?
- Newton's bucket and Mach's principle
- Uncertainty
- Super symmetry and the "Sparticles"

- The beginning of everything
- The double slit experiment results

Ultimately, things should get simpler as we understand more and more about the universe and we may be at that point now, when the complexities, some of which we have created, finally start to resolve into a unified understanding.

Our ultimate quest is for the unification of all aspects of the physical world, a common, single cause for all effects, known as the "Grand Unification". The idea of time as a wavelike field of energy, the source of all energy, gives us this single cause and is surely the only possible basis for the actual "Theory of Everything".

Potentially, with the time curvature of gravity at the large macro scale, the wave nature of time in its own field, the make up of matter at particle scales and the quantum nature of time itself at the Planck scale, we have an explanation for everything in terms of quantum energy, the energy of the Planck time.

None of these ideas conflict with accepted science, or experimental results. Indeed they explain certain, present day problems in physics and we now have a rigorous causality for some previously inexplicable effects like inertial time dilation and the wave nature of light.

Certainly, the binary universe with a net zero energy, a null creation, a temporal explanation for everything, the *standing* wave of time and an eternal, perfect symmetry, is certainly beguiling, possibly compelling.

The potential benefits from this theory are enormous and will open up many new avenues for scientific progress, yet the costs of the proposed experiments to verify it are quite modest. It would therefore be the utmost folly if we were to neglect to execute these experiments without delay and in particular, the test of the relative *blue shift* observed from the more accelerated frame. The likely results from this experiment will turn the world of science upside down

This book is not a peer reviewed scientific paper, it has not been checked by anyone other than myself and it is not scientifically rigorous in all aspects. It is, in some respects, naïve when compared to Quantum Mechanics and Quantum Field

Theory and I bow to the great work being done today by physicists around the world. I did say in my introduction that I am not a physicist, but an Engineer with the talent to view the whole and to see how everything fits together. This is what I have done and as a result it has become as clear as day to me that the B.U.T. *must* be true. This is one of those times when you consider a new idea and everything falls into place and that alone tells us it is highly likely to be correct.

From a statistical analysis alone, if it were false, it is highly improbable that the B.U.T. would fit all known science and also explain many present-day conundrums. If the theory were flawed, then it would surely have failed against at least one and probably many, aspects of accepted science, but it has not.

It fits accepted science perfectly, although not the mainstream *interpretations* of all aspects. But these interpretations are not science, they are opinions. It goes on to answer questions that have been left, ignored and unasked for many years, as well as questions currently posed, as yet unanswered. Finally, it presents much better explanations for some issues that have been dismissed with speculative, unscientific and unproven justifications. I therefore put it to you that the B.U.T. has much merit and as such, it surely must be pursued. My hope is, having read this book, your mind has been opened to this new way of thinking, that there is only the one fundamental in the universe, the wavelike field of energy we know as time. It is not surprising therefore that everything can be explained from a temporal aspect and that everything has a wave function.

This may be difficult for some to come to terms with, and that is understandable. We are like fish in the sea wondering why we bob up and down, not knowing we live in water and it is the water which is wavelike. We believe we move in the one direction of the current because it is just the way we see things, just an attribute of our perceptions. This one way motion cannot possibly be a real, physical phenomenon can it? We have to use our imaginations in order to overcome this mistaken belief since no mathematical workings can expose it for the illusion that it is. Today's math works perfectly well for any view of time, especially if we make it fit, but only the B.U.T can explain it all in real, physical terms. I am certain these ideas will be criticised because they lack mathematical proof, but the mathematics of accepted theory supports both options and it is unscientific to believe in either, simply because one was first believed, before the other. In this situation, it is not any mathematical workings that will demonstrate which is correct, but only logical deduction applied to our current interpretations. The logical processes applied in this book have cut through the mistaken beliefs of the mainstream like a hot wire through butter.

As I said in my introduction, everything in this world has to be a physical entity or process and there is nothing beyond the physical, the only exception being the fundamental energy quanta as I discovered in the process of my writing. The progression of time (above the scale of the wave), must also be a physical process and to believe differently is self-deception, quite literally on a cosmic scale, a stubborn, irrational belief in an infinitely available process of time which is not substantiated.

The physical nature of time has been staring us in the face now for many decades, yet we have neglected to investigate it despite all the clues and pointers prompting us to think along these lines. This issue is of such monumental importance that we *must* now consider it from a physical aspect. We *must* consider the Binary Universe Theory.

We must not only realise that time is a physical phenomenon or process, but its energy is finite and when it runs out, it really does cease in absolute terms. Everyone, including the most prominent scientists, seem to believe that time will just carry on as normal for someone at the speed of light or on crossing the event horizon of black hole. There is no scientific justification for this belief in some magical, persistently available, passage of time, irrespective of circumstance.

This mistaken belief is born of our personal experience and we insist on believing in our continuing progression through time, our continuing changing of state, whatever our frame of reference. Certainly, at the speed of light, or on the event horizon of a black hole, we would not perceive that time had stopped because our brain synapses would be unable to send even one signal for any mental process of recognition. So, we have been stuck with this preconception that we must continue to change, even when we no longer *can* change.

Time, like everything else *must* be a finite, physical process. Indeed, physicists do tell us it passes at the rate of 1.855×10^{43} Planck times per second and this is clearly a finite rate. So, when it slows down, relative to some other frame of reference, it makes no sense to then claim that this effect is purely relative. If the rate of time is finite, then it must eventually slow to a zero rate in extreme circumstances.

This process of change cannot remain available to anyone in these situations. Time dilation is relative certainly, but it is also, *absolute*.

The passage of time is a finite, physical process

The Binary Universe Theory is internally consistent. It is consistent between all aspects of itself and it is consistent with accepted science, despite the differences of opinion I have pointed out. More than this, the B.U.T. provides intuitive explanations for conundrums which the mainstream cannot yet explain.

I hope that just one of the physicists amongst you might see the value of this approach and take up the cause, or at least bring pressure to bear to execute the proposed experiment. Even if you strongly disagree, I hope you might still do this, if only to shut me up, to silence this heretical train of thought.

If I am right and the results from the experiments do confirm the theory, then the basis of physics for the twenty first century will change from that moment onwards. If I am wrong, well, I hope I have encouraged you to think about science a little differently, at least for a short while.

Except...

...I am not wrong!

"Behind it all is surely an idea so simple, so beautiful, that when we grasp it - in a decade, a century, or a millennium - we will all say to each other, how could it have been otherwise?"

John Archibald Wheeler

About the author

Ken Hughes is a professionally qualified Mechanical Engineer, now retired. Born in Denton, Lancashire, England in 1947 during the period of austerity immediately after the Second World War, he is a native citizen of the United Kingdom of Great Britain.

Of northern stock, he inherits the down to earth attitudes typical of that region. As a young child, he played in the rubble that was the city of Portsmouth and throughout his life has borne witness to the rebuilding and regrowth of the UK. He was educated in the local schools and colleges and attended the University of Portsmouth where he read Mechanical Engineering.

He has spent the majority of his career in technical and managerial roles in the process plant engineering and construction industry. This has been mainly within the UK, but with some time spent in the USA in the 1980s and more recently on continental Europe, as a freelance Engineering Consultant.

He has always had a marked interest in physics, astronomy and cosmology and a fascination for gravity, but his professional engineering career has not utilised his scientific leanings and abilities. So, perhaps later in life than might have been expected, he has finally found the time, the energy and the motivation to finally clarify these outstanding issues in science that have been troubling him over the years.

Index

E

Earth, 33, 34, 36, 38, 42, 64, 73, 78, 79, 105, 113, 115, 117, 120, 124, 127, 129, 130, 132, 133, 134, 135, 137, 138, 142, 147, 148, 152, 153, 154
Einstein, 26, 30, 42, 43, 44, 46, 53, 54, 61, 74, 119, 135, 180, 182, 184, 185
Electric fields, 82, 122
Electromagnetic radiation, 56, 186
Electron, 121, 122, 124, 127, 129
Electroweak, 30
Empirical methods, 32
Energy, 42, 47, 48, 49, 61, 62, 69, 78, 82, 121, 122, 124, 134, 177, 178, 179, 186, 214
Entity, 48, 53, 55, 57, 60, 61, 68, 72, 74, 83, 90, 92, 97, 98, 117, 118, 120, 121, 133, 134, 142, 148, 177, 181, 184, 186, 335
Equilibrium point, 113
Escalators, 105
Escape velocity, 76
Event horizon, 76, 77, 138
Events, 58, 63, 70, 71, 77, 80, 83, 97, 99, 138, 208

F

Faster clock, 99
Field strength, 82
Finite speed, 95
Flat space, 46, 54, 58, 111, 112, 151, 182
Flat space time, 111
Flat space-time, 46
Flatlanders, 70
Fourth dimension, 60
Frame of reference, 56, 61, 62, 65, 69, 72, 78, 79, 105, 121, 122, 130, 186, 187
Free fall, 113, 134, 135, 140, 178, 180, 182, 183
Frequency, 187, 188
Frozen in time, 95

G

g, 122, 137, 140, 144, 152, 153, 186
Galileo Galilee, 34
Gaussian coordinate system, 71
Geocentric, 33
Geodesics, 45
Geometric solutions, 72
Geometry, 62, 64, 71, 79, 81, 91, 92, 107
Global positioning system, 72
Gravitation, 27, 28, 30, 34, 35, 41, 43, 45, 46, 54, 98, 109, 111, 112, 113, 117, 118, 119, 124, 125, 130, 133, 134, 137, 145, 146, 147, 148, 149, 151, 176, 177, 178, 179, 182, 194, 336
Gravitational acceleration, 29, 45, 49, 112, 117, 118, 122, 132, 134, 135, 136, 137, 140, 142, 144, 145, 146, 148, 149, 175, 180, 181, 335
Gravitational field, 44, 45, 47, 48, 52, 53, 56, 57, 58, 65, 68, 69, 76, 78, 112, 117, 118, 119, 124, 132, 134, 135, 138, 140, 142, 151, 175, 176, 177, 178, 180, 181, 182, 183, 186, 335
Gravitational mass, 43
Gravitational potential, 49, 108, 177
Gravitational shielding, 115, 119, 147
Graviton, 29, 44, 335
Gravity, 27, 28, 29, 30, 41, 42, 43, 44, 45, 47, 48, 49, 54, 76, 99, 111, 112, 113, 117, 118, 120, 132, 134, 135, 140, 145, 146, 151, 152, 177, 178, 181, 336

H

Hafele and Keating, 73
Hammer and feather, 43, 180
Heliocentric, 34, 35
Heliocentric system, 34
Huygens, 37, 38
Hypothesis, 148
Hypothetical, 64, 119, 138

I

Illusion, 185
Image, 77, 95
Increment, 92, 132

V

Vacuum, 37, 47, 53, 57, 61, 83, 97, 117, 142, 148, 180, *distortion of a vacuum*

Vector, 111, 120, 132, 151, 182, 187

Velocity, 38, 62, 69, 76, 81, 82, 83, 86, 87, 88, 92, 93, 94, 95, 96, 97, 98, 99, 105, 130, 132, 134, 179, 181, 183, 184, 208

Visualise, 37, 54, 109, 115, 121, 124, 181

Void, 37, 60, 68

W

Wave / particle duality, 181

Wave form, 37, 187, 189, 214

Wave like, 214

Wave nature, 37, 183, 184, 187

Wave packet, 121

Wave-particle duality of light, 48

Waves, 37, 38, 48, 214

World Lines, 45

Y

Young and Fresnel, 38

Z

Zero, 38, 62, 68, 95, 113, 121, 130, 134, 152, 178, 179, 181, 187, 208

www.ingramcontent.com/pod-product-compliance
Lightning Source LLC
Chambersburg PA
CBHW050634190326
41458CB00008B/2262